Differentiated Resourceful Utilization of Rural Organic Wastes

Mingxiao Li · Xuan Jia · Beidou Xi ·
Jiaqi Hou · Dongming Liu · Yan Hao

Differentiated Resourceful Utilization of Rural Organic Wastes

Springer

Mingxiao Li
Chinese Research Academy
of Environmental Sciences
Beijing, China

Xuan Jia
Beijing Technology
and Business University
Beijing, China

Beidou Xi
Chinese Research Academy
of Environmental Sciences
Beijing, China

Jiaqi Hou
Chinese Research Academy
of Environmental Sciences
Beijing, China

Dongming Liu
Chinese Research Academy
of Environmental Sciences
Beijing, China

Yan Hao
Chinese Research Academy
of Environmental Sciences
Beijing, China

ISBN 978-981-15-2711-1 ISBN 978-981-15-2712-8 (eBook)
https://doi.org/10.1007/978-981-15-2712-8

This Springer imprint is published by the registered company Springer Nature Singapore Pte Ltd.
The registered company address is: 152 Beach Road, #21-01/04 Gateway East, Singapore 189721, Singapore

Preface

China is the largest producer of organic wastes in villages and towns in the world. In accordance with data released by the Ministry of Environmental Protection of the People's Republic of China, the pollution from agricultural source accounts for $1/3 \sim 1/2$ of the total pollution in China, so it has become an important source of water, soil and air pollution in rural areas. The resourceful utilization of organic wastes in villages and towns in China is challenged by the difficulties including the lack of applicable technologies and equipment. Based on this, research and development as well as optimization and upgrading of key technologies and equipment, such as energy, fertilization, forage, and materialization are carried out, so as to realize the unification of resource technologies and equipment. Aiming at this, this book systematically introduces the research on technologies for differentiated resourceful utilization of organic wastes of villages and towns. Around the main clue of the differentiated resourceful utilization of organic wastes in villages and towns, this book focuses on general thinking of the identification of source characteristics of organic wastes, integration of generic technologies, development of key technologies and research on management modes.

In this book, we (1) introduced the current situation of organic wastes in villages and towns in China and the necessity of Research on key technologies and practical equipment for differentiated resourceful utilization of organic wastes, (2) gave some high-efficient anaerobic fermentation technology including pre-treatment and evaluation methods on fermenting materials and the energy they produced, as well as the low-cost, energy-saving equipment, (3) proposed a series of biological aerobic treatment methods, including preparation of bacteria agent, method of humification, control of secondary pollution, quality evaluation of composting products, research and development of composting equipment, etc, (4) also introduced the development of the standard system for differentiated resourceful utilization of organic wastes in villages and towns, So that readers not only understand the treatment technology, but also understand its treatment standards and policies, (5) finally provided some projects of rural waste treatment based on the policies, standards and technical methods introduced in previous contents.

The content of this book includes the current status, technologies, policies and standards of organic waste disposal in villages and towns in China as well as engineering cases. It can be used as a reference for workers, technicians and rural living environment managers in the waste treatment industry.

Due to the limited level of editors, there are unavoidable omissions in the book, so please oblige us with your valuable comments.

Beijing, China

Mingxiao Li
Xuan Jia
Beidou Xi
Jiaqi Hou
Dongming Liu
Yan Hao

Contents

Abbreviations

AGE	Agarose gel electrophoresis
BES	2-bromoethanesulfonate
C/N	Carbon-nitrogen
CHMP-AF	Combined production of hydrogen and methane via anaerobic fermentation
COD	Chemical oxygen demand
CPC	National Congress of the Communist Party of China
CSTR	Continuous stirred tank reactor
DOM	Dissolved organic matter
EEM	Excitation-emission matrix
EPA	Environmental protection agency
Ex/Em	Excitation/Emission
FRI	Fluorescence regional integration
FVWs	Fruit–vegetable wastes
GC-MS	Gas chromatography-mass spectrometry
GI	Germination index
HULIS-C	Carbon in humic-like substances
LCFAs	Long-chain fatty acids
PADRs	Piggery anaerobic digested residues
PARAFAC	Blind parallel factor
PCR	Polymerase chain reaction
PCS	Peptone cellulose solution
PLC	Programmable logic controller
PPP	Public private partnership
sCOD	Soluble chemical oxygen demand
SEM	Scanning electron microscope
TCA	Tricarboxylic acid
TN	Total nitrogen
TOC	Total organic carbon
TS	Total solid

VFA	Volatile fatty acid
VOCs	Volatile organic compounds
VS	Volatile solid

Chapter 1
Introduction

Abstract According to statistical bulletin of urban and rural construction, there were 20,883 towns and 10,872 villages in total in China and the total rural population was 9.58×10^8 by the end of 2016. The total investment in the construction of villages and towns in China was 1.5908×10^{12} yuan, while the investment in environmental health was 4.23×10^{10} yuan, accounting for 10.5% of the total investment in construction of municipal utilities. At the end of 2016, water supply in built-up area of towns was 83.86% and daily water consumption per capita was 99.01 L. Fuel gases were supplied for 49.52% of the population and road area per capita was 12.84 m^2. Moreover, density of culverts in drainage pipelines was 6.28 km/km^2 and park and green area per capita reached 2.46 m^2. Centralized water supply was realized in 68.7% of the administrative villages in China and sanitary sewage and household wastes are separately disposed in 20 and 65% of administrative villages. The infrastructure construction and quality of living environment in villages and towns have been significantly improved. However, the 13th Five-Year Plan for Comprehensive Improvement of Rural Environment of the People's Republic of China shows that 40% of villages still lack of wastes collection and disposal facilities and have prominent problem of dirty, disorderly and bad environment caused by waste pollution in villages and towns. Therefore, it is urgent to take effective measures to make up the shortcoming of controlling waste pollution in villages and towns.

1.1 Current Situations of Construction of Villages and Towns and Environmental Protection in China

According to statistical bulletin of urban and rural construction, there were 20,883 towns and 10,872 villages in total in China and the total rural population was 9.58×10^8 by the end of 2016. The total investment in the construction of villages and towns in China was 1.5908×10^{12} yuan, while the investment in environmental health was 4.23×10^{10} yuan, accounting for 10.5% of the total investment in construction of municipal utilities. At the end of 2016, water supply in built-up area of towns was 83.86% and daily water consumption per capita was 99.01 L. Fuel gases were supplied for 49.52% of the population and road area per capita was 12.84 m^2.

M. Li et al., *Differentiated Resourceful Utilization of Rural Organic Wastes*,
https://doi.org/10.1007/978-981-15-2712-8_1

Moreover, density of culverts in drainage pipelines was 6.28 km/km^2 and park and green area per capita reached 2.46 m^2. Centralized water supply was realized in 68.7% of the administrative villages in China and sanitary sewage and household wastes are separately disposed in 20 and 65% of administrative villages. The infrastructure construction and quality of living environment in villages and towns have been significantly improved. However, the 13th Five-Year Plan for Comprehensive Improvement of Rural Environment of the People's Republic of China shows that 40% of villages still lack of wastes collection and disposal facilities and have prominent problem of dirty, disorderly and bad environment caused by waste pollution in villages and towns. Therefore, it is urgent to take effective measures to make up the shortcoming of controlling waste pollution in villages and towns.

China is the largest producer of organic wastes in villages and towns in the world. According to statistics, annual discharge of livestock manures is 2.61×10^9 tons, along with 7.0×10^8 tons of crop straws, 2.5×10^4 tons of plastics including abandoned agricultural films, and 1×10^8–1.5×10^8 tons of vegetable wastes. In accordance with data released by the Ministry of Environmental Protection of the People's Republic of China, the pollution from agricultural source accounts for 1/3–1/2 of the total pollution in China, so it has become an important source of water, soil and air pollution in rural areas. China, as a country with the greatest capacity of breeding livestock and poultry in the world, has more than 8×10^4 large-scale farms with more than ten thousand livestock and poultries. Livestock and poultries in these farms discharge over 2×10^9 tons of excrements each year, which seriously damages farm lands and underground water. Agricultural waste resources are very rich and have huge production in China. They are mainly distributed in major grain producing areas, such as Hebei, Inner Mongolia Autonomous Region, Liaoning, Heilongjiang, Shandong, Hubei, Hunan, Jiangxi, Anhui, Sichuan and Yunnan. Based on the calculation of agricultural output in China, totally about 7×10^8 tons of straw resources are produced in each year, in which about 5×10^8 tons can be collected, while the utilization rate is only 33%. With the sustained and rapid development of Chinese economy and the continuous promotion of new urbanization construction, the industrial pattern and consumption mode of villages and towns have undergone significant changes. However, treatment and disposal techniques for organic wastes in villages and towns are still in the initial stage, so facilities for collecting and disposing wastes are inadequate. Moreover, a lot of wastes are piled up in open air or simply filled, which causes secondary pollutions of a lot of leachate and malodorous gas, wastes resources and breeds germs, greatly harming human healthy. At present, rural ecological environment has become a prominent shortcoming in building a moderately prosperous society in all respects. To promote ecological civilization, build beautiful villages and resolutely fight against ecological environmental pollution, the control of agricultural non-point source pollution is the key.

Since implementing the policy of "award for promoting governance" in 2008, China has issued more than 10 normative documents and allocated 3.15×10^{10} yuan of special funds for rural environmental protection. Jiangsu province and the Ningxia Hui Autonomous Region have been selected as pilots of large-scale continuous environment improvement covering the whole province, so as to guide and promote continuous improvement of rural environment in the whole country. In 14,000 monitoring sites in 700 counties across China, rural environmental health, involving the harmlessness treatment of rural sewage, wastes, and excrements, the soil hygiene and the vector control has been monitored. By the end of 2014, 5.9×10^4 villages had carried comprehensive environmental improvement, directly benefiting more than 1.1×10^8 people, so comprehensive improvement of rural environment has achieved remarkable results. Since the 18th National Congress of the Communist Party of China (CPC), General Secretary of the Central Committee of CPC Xi Jinping has attached great importance to ecological civilization construction and rural environmental protection and put forward a series of new ideas, judgments and requirements on building ecological civilization and strengthening environmental protection. He stressed that if China wants to be a beautiful country, the countryside must be beautiful, and a beautiful country relies on beautiful countryside. China will continue to promote the construction of new socialist countryside and build a happy homeland for farmers. China should pay attention to ecological and environmental protection in new rural construction, carry out comprehensive improvement of the rural living environment according to local conditions, change the dirty, messy and poor conditions in rural areas as soon as possible, and give farmers a clean and tidy living environment. In February, 2018, the State Council issued No. 1 document of the central government, namely, the Opinions of the CPC Central Committee and the State Council on Implementing the Strategy of Rural Revitalization. It is proposed to strengthen the prevention and control of agricultural non-point source pollution, carry out green development in agriculture and realize input reduction, cleaner production, waste resource recovery and ecological industrial mode. Obviously, strengthening rural environmental protection is an inevitable requirement for building ecological civilization, an important task of balancing urban and rural development, and an urgent need to improve and guarantee people's livelihood. To build a moderately prosperous society in all respects, rural areas are the top priority, while the environment is the key. Rural environment is not only the weakest point, but also the most potential breakthrough point and innovation point.

1.2 Necessity of Differentiated Resourceful Utilization of Organic Wastes in Villages and Towns

With the constant and rapid development of economy, villages and towns in China have rapidly entered the society of commodity of modern market economy from the self-supporting agricultural society. Household goods in villages and towns have been separated from most local production activities. Planting has been almost completely separated from breeding and the closed-circuit production mode of organic matter in the production and life of villages and towns has been completely cut off by modern agriculture and sustainable commercialization, mechanization and scale modes. Subsequently, the output of a large number of organic wastes in villages and towns increases sharply, while relevant treatment and disposal technologies and supporting facilities are relatively backward, which seriously influences the quality of ecological and living environment of villages and towns. China should develop the mode with Chinese characteristics for resourceful utilization of organic wastes, break through restriction of diversity and versatility of sources, and different natural conditions of organic wastes and address ecological and health risks associated with the reuse of toxic pollutants in wastes. These are not only the needs for continuous improvement of environmental quality, but also objective demands for protecting drinking water safety, reducing discharge of major pollutants in rural areas and improving the quality and safety guarantee of agricultural products. Differentiated resourceful utilization of organic wastes in villages and towns mainly aims at organic wastes generated from the decentralized breeding industry, living source, tourism industry and farmers' markets in villages and towns. Based on this, research and development as well as optimization and upgrading of key technologies and equipment, such as energy, fertilization, forage, and materialization are carried out, so as to realize the unification of resource technologies and equipment.

(1) Research on key technologies and practical equipment for differentiated resourceful utilization of organic wastes in villages and towns is an objective requirement for solving environmental problems caused by the pollution of rural organic wastes and improving social development and people's livelihood.

    Organic wastes in villages and towns in China have four pollution characteristics, i.e. large quantity, poor quality, low price and many hazards. Hundreds of millions of wastes are produced every year. Owing to the content of beneficial components and grade of available objects are low, while the content of harmful components is high, it costs high for disposing agricultural wastes. At present, the utilization rate of organic wastes in villages and towns in China has not been improved obviously. Straw burning and randomly discard or discharge of livestock manures caused by intensive cultivation result in that environmental pollution has become increasingly serious. This turns the original resources into pollution sources, which greatly influences ecological environment. In view of organic wastes from decentralized breeding industry, living source, tourism industry and farmers' markets in villages and towns, it is extremely urgent to carry out researches on key technologies for differentiated resourceful utilization of organic wastes in villages and towns and industrialization promotion.

(2) Carrying out research on key technologies and practical equipment for differentiated resourceful utilization of organic wastes in villages and towns to solve rural energy shortage and protect ecological environment driven by technology and equipment innovation.

    With the rapid development of industry and agriculture in villages and towns in China, pollution is also becoming serious. In particular, the development of agricultural production and the changes of rural living and production habits make various organic wastes in villages and towns become the important rural pollution sources. In the meanwhile, as long as the resources are recovered by using reasonable methods, this can relief energy pressures in villages and towns in China, save natural resources, protect ecological environment and unify ecological, economic and social benefits. At present, China has done a lot of work on resourceful utilization of rural organic wastes, mainly including fertilization, forage and fuelization of organic wastes. However, due to the lack of new technologies and modes, these problems, such as low utilization value of resourceful products and low quality of products, like fertilizer and feed are widespread. Particularly, with the increasing shortage of natural resources and dramatic increase of wastes, more and more attention has been paid on organic wastes in villages and towns. In nowadays with market economy and industrialized operation, it is urgent to develop diversified technologies and equipment for differentiated resourceful disposal of organic wastes.

(3) Research on key technologies and practical equipment for differentiated resourceful utilization of organic wastes in villages and towns is a strong guarantee to improve the resourceful efficiency of organic wastes.

In China, there are obvious spatial differences in the geographical characteristics of villages and towns and imbalanced social and economic development. Moreover, different regions have greatly different industrial structure, ecological carrying capacity and environmental pressure. For these reasons, the output and compositions of organic wastes in villages and towns in different regions are quite different and the unified resourceful technical mode is obviously not suitable for the disposal of organic wastes in villages and towns showing regional differences. In addition, the existing utilization technologies for organic wastes in villages and towns are immature and have poor stability and high cost for resourceful utilization, which results in different effects of resourceful utilization of organic wastes in villages and towns in improving economic efficiency and reducing farmers' labor intensity. The major technologies for resourceful utilization of organic wastes in villages and towns mainly concentrate in large and medium-sized livestock and poultry breeding enterprises, while the researches on small-scale, resourceful and applicable technologies and equipment that suit for the majority of villages and towns are rare. Therefore, it is urgent to carry out research and development on resourceful disposal mode and complete technologies and equipment for organic wastes of villages and towns in different regions.

(4) Research on key technologies and practical equipment for differentiated resourceful utilization of organic wastes in villages and towns is important content of building resource-conserving society in villages and towns.

The construction of conservation-minded society needs the joint efforts from all fields of society, and resourceful utilization of organic wastes in villages and towns is important content for building conservation society. Organic wastes in villages and towns, such as livestock manures, contain a large amount of plant nutrients (like organic matter, nitrogen and phosphorus). By using these organic wastes, high-quality organic fertilizer and commercial organic compound fertilizer can be produced by utilizing the preparation technology of organic fertilizer and organic-inorganic compound fertilizer. Treating organic wastes by using biogas fermentation technique can produce biogases and biogas manure. The recycling of matter and energy can also solve contradiction of fuel, fertilizer and feed and increase farmers' income.

1.3 Research Status of Resourceful Utilization of Organic Wastes in Villages and Towns

Organic wastes in villages and towns in China demonstrate four major pollution characteristics, i.e. large quantity, poor quality, low price and much harm. In view of the characteristics of organic wastes in villages and towns in different regions, the multiple technical system and comprehensive management mode for differentiated resourceful utilization of organic wastes are built according to local conditions. The aim is to provide technical supports for resourceful utilization of organic wastes in villages and towns in different regions in China. Based on the current situations of popularization of biological fermentation and biogas projects of organic wastes in villages and towns, environmental safety of resourceful utilization of secondary products is improved through research and development of key techniques and upgrading of the existing techniques. This forms the technical system of multistage utilization of organic wastes in villages and towns. In accordance with the production situations of organic wastes from tourism in villages and towns, the integrated mode of waste disposal and tourism landscape was proposed according to local conditions, which gives new connotation to village and town tourism. In view of characteristics of organic wastes from pedlars' markets in villages and towns, the differentiated resourceful utilization mode with optimized classification and collection was put forward, which provides a model for resourceful disposal of organic wastes from pedlars' markets in villages and towns.

The resourceful utilization of organic wastes in villages and towns in China is challenged by the difficulties including the lack of applicable technologies and equipment. Aiming at this, this book systematically introduces the research on technologies for differentiated resourceful utilization of organic wastes of villages and towns. Around the main clue of the differentiated resourceful utilization of organic wastes in villages and towns, this book focuses on general thinking of the identification of source characteristics of organic wastes, integration of generic technologies, development of key technologies and research on management modes. Based on this, organic wastes in villages and towns, the organic wastes are classified as scattered breeding ones, mixed wastes from courtyard, mixed wastes from pedlars' market and tourist catering wastes in view of the sources. In accordance with characteristics and scales of organic wastes in different villages and towns, the key technologies and practical equipment for differentiated resourceful utilization of organic wastes in villages and towns in China were deeply researched. The purpose is to optimize and upgrade the key technologies, such as high-efficiency energy regeneration and fertilization suiting for being promoted in towns and villages, research a complete set of diversified equipment, and compile technical standards for quality-based resourceful utilization of organic wastes in villages and towns. In addition, it aims to establish a base of research and development of technologies and equipment for resourceful utilization of organic wastes in villages and towns and their industrialization. It also attempts to promote the manageable technical mode of low pollution, cost, and maintenance requirement suiting for differentiated resourceful disposal of organic wastes in villages and towns in China, to provide technical supports for improving resourceful and harmless disposal level of organic wastes in villages and towns.

Chapter 2
High-Efficient Anaerobic Fermentation Technology of Organic Wastes in Villages and Towns

Abstract The accumulation of organic waste is thought to be critical levels in the world. The organic wastes require to be managed in a sustainable way to avoid depletion of natural resources, minimize risk to human health, reduce environmental burdens and maintain an overall balance in the ecosystem. Anaerobic fermentation technology is currently applied to the treatment organic wastes and realize energy recovery. The chapter focuses on process system, influence factors, pre-treatment technologies of organic wastes and high-efficient anaerobic fermentation technology. In addition, the chapter reviewed energy recovery efficiency and anaerobic fermentation equipment. It provides a broad overview of the digestibility and energy production (biogas) yield of a range of substrates of organic wastes in villages and towns. The involvement of a diverse array of microorganisms and metabolites and environmental factors on the efficiency of the process has been comprehensively addressed to obtain higher efficient anaerobic fermentation performance.

2.1 Overview

Since stepping into the twenty-first century, under dual pressures of energy and crisis mitigation, more and more attention has been paid on renewable and low-pollution biomass energy in the world. Owing to anaerobic fermentation process has characteristics, such as no need for oxygen supply, being capable of saving energy and reducing cost, significant comprehensive benefits and suiting for villages and towns, it has become a research hotspot of industrial application. Since the 1970s, China has vigorously developed rural biogas projects and the amount of household biogas digesters has reached 5.25×10^6 during the Ninth Five-Year Plan period. Moreover, 633 large and medium-sized biogas projects were built to intensively supply biogas for 8.4×10^4 families. The construction of rural household biogas projects has effectively utilized rural living and production wastes, improved the living environment of rural residents, eased the pressure of rural energy demands, and contributed to the emission reduction of CO_2 and SO_2. However, the construction of biogas projects in villages and towns still needs to further improve supporting

M. Li et al., *Differentiated Resourceful Utilization of Rural Organic Wastes*,
https://doi.org/10.1007/978-981-15-2712-8_2

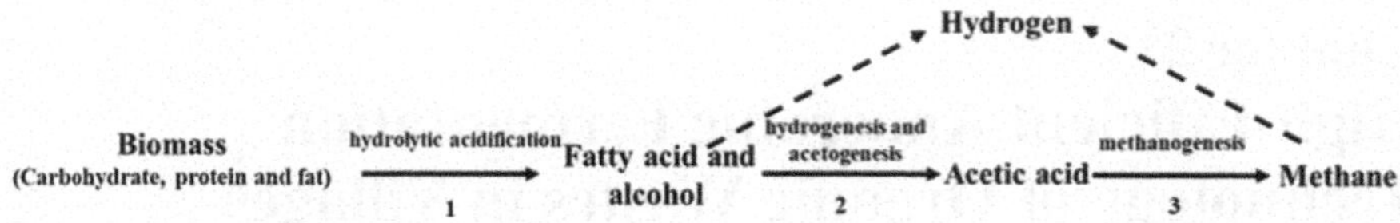

Fig. 2.1 Conceptual model of anaerobic fermentation process

facilities, optimize technological process, raise the comprehensive utilization level of energy and improve the service system.

2.1.1 Traditional Anaerobic Fermentation Process

Anaerobic fermentation is a process to decompose organic matter and produce H_2, CH_4 and CO_2 under the joint efforts of anaerobic or facultative anaerobic microorganisms under anaerobic conditions. Its fundamental differences with aerobic treatment process lie in that it uses combined oxygen, carbon, sulfur and nitrogen rather than molecular oxygen as hydrogen acceptors. This technology can realize harmlessness, reduction, stabilization and resourceful utilization of organic wastes and shows advantages, like low cost, high efficiency and energy saving, so it is considered as a sustainable treatment technology.

The traditional anaerobic fermentation process is shown in Fig. 2.1. Step 1 demonstrates hydrolysis and fermentation, that is, macromolecular substances, such as carbohydrate, protein and fat are hydrolysed into monosome by extracellular enzyme and produce multiple intermediate products. Step 2 displays hydrogenesis and acetogenesis. Various products of hydrolysis and fermentation are continuously decomposed into methanogenic precursors, such as acetic acid and hydrogen. Moreover, step 3 shows methanogenesis. The methanogenic precursors are transformed into methane, so as to completely remove organic pollutants.

2.1.2 Two-Phase Anaerobic Fermentation Process

In the early period of the 1970s, Pohland and Borchardt firstly proposed the concept of phase separation in the field of anaerobic digestion. The most significant characteristic of the two-phase anaerobic system is to realize phase separation, so that all kinds of bacteria participating in anaerobic fermentation are optically distributed in two-phase anaerobic bio-treatment system, thus improving treatment effects of anaerobic fermentation process. However, for a long time, the two-phase anaerobic process researched in the world is mainly used as a treatment system of organic pollutants, while less attention is paid on its effects while being utilized as an energy recovery process. The research demonstrates that by taking glucose as an example, when acetic acid is the only intermediate metabolite, the maximum theoretical hydrogen production rate of 1 mol of glucose is 4 mol. However, the actual hydrogen production rate is far smaller than the theoretical value, generally being 0.5–2.5 mol and the intermediate metabolite is generally a mixture containing volatile fatty acid (VFA) and alcohol, while they are the best substrate of methanogens. The highest theoretical energy recovery efficiency of hydrogen production in biochemical reaction is only 33.5% through anaerobic fermentation, which is lower than 83.2% of that of methane production via anaerobic fermentation. Furthermore, the theoretical energy recovery efficiency in the combined production of hydrogen and methane via anaerobic fermentation (CHMP-AF) can reach 89%. Therefore, coupling the CHMP-AF process under phase separation can obtain two energies, i.e. hydrogen and methane. This effectively improves energy recovery rate of organic wastes through anaerobic fermentation and overcomes feedback inhibition of acidic metabolite and hydrogen partial pressure in traditional processes, which is favorable for long-term and stable operation of the process.

2.1.3 Influence Factors of Anaerobic Fermentation

2.1.3.1 Temperature

Temperature is an important factor influencing biogas production efficiency in anaerobic fermentation. The temperature mainly affects hydrolysis degree of materials, enzyme activity and structure and function of microbial population in the fermentation. According to the influences of temperature on microbial activity, the anaerobic fermentation can be divided into four temperature ranges, i.e. low-temperature fermentation at 15–20 °C, moderate-temperature fermentation at 30–35 °C, high-temperature fermentation at 50–55 °C and ultra-high-temperature fermentation above 70 °C. Of them, anaerobic fermentation at moderate and high temperature is mainly studied, because too low temperature can slow down metabolic rate of microorganisms, while too high temperature can inactivate protease in cells. Both of the two conditions probably reduce biogas production efficiency.

2.1.3.2 pH Value

pH value is an important factor for measuring stable operation of the anaerobic fermentation system. It mainly affects the changes of enzyme activity in metabolic process. Through the changes of membrane charge, it has effects on transmission of substances on the two sides of cytomembrane of microorganisms and adsorption of nutrients. Moreover, different pH ranges correspond to different fermentation types for hydrogen production, thus affecting biogas production efficiency. It shows impacts on succession of structure of microbial community. The vast majority of microorganisms grow in the pH range of 5–9. Acid-producing bacteria are suitable in the pH range of 5–8, while methanogens suits for the range of 6.8–7.2 of pH value.

2.1.3.3 Carbon-Nitrogen Ratio

Carbon-nitrogen (C/N) ratio, as a key factor affecting anaerobic fermentation, mainly influences anabolic process of microorganism itself and oxidation process of organic matter in organisms and participates in the energy-releasing metabolic process of bacteria. A too high C/N ratio leads to inadequate nitrogen content and low buffer ability of fermentation liquids, so that the liquids are easily acidified. A too low C/N ratio implies the presence of excess carbon content, which easily results in accumulation of ammonia nitrogen, so that anaerobic fermentation process is inhibited. An appropriate C/N ratio is beneficial for stable operation of anaerobic fermentation. It is generally considered that the suitable C/N ratio for anaerobic fermentation is in the range of 20–30, while C/N ratio in the range of 15–20 in the actual operation can guarantee high-efficient utilization of organic substrates.

2.1.3.4 VFAs and Ammonia Nitrogen

VFAs refer to micromolecular acid substances containing less than six carbon atoms formed in hydrolytic acidification and they belong to important substrate materials in the methanogenic process. They mainly include substances, such as acetic acid, propionic acid, butyric acid and a small amount of pentanoic acid. High-concentration VFAs can reduce pH, inhibit activity of methanogens and acidify and thus collapse the system. Ammonia nitrogen produced by decomposing substances, like protein and amino acids based on catabolism of microorganisms has positive effects on maintaining stable pH value of fermentation liquids. Ammonium ion (NH_4^+) and free ammonia (NH_3) are the dominant forms existing in the transformation of organic nitrogen, while free ammonia can influence the balance of positive ions in cells by passing through cytomembrane, thus producing toxicity inhibition in anaerobic fermentation. VFAs and ammonia nitrogen in the anaerobic fermentation system can form a buffer system, so that fermentation system has certain buffer ability, which shows important regulations on long-term and stable operation of anaerobic fermentation.

2.1.3.5 Microbial Community

The essence of anaerobic fermentation is to decompose organic matter through the effects of microorganisms and the type, quantity, synergy and antagonism of microbial community directly affect the operation state of anaerobic fermentation. Different fermentation stages present different dominant bacterial communities. In the acidification stage, hydrogen-producing bacterial community, hydrogen-consumption bacterial community and a kind of microorganisms showing a competitive relationship with hydrogen-producing bacterial community are dominant. In the methanogenic stage, bacterial communities are dominated by methanogens and bacteria producing hydrogen and acetic acid. These bacteria guarantee the subsequent stable methanation reaction through balanced adjustment of hydrogen partial pressures. The changes of compositions of microbial community in anaerobic fermentation directly affect metabolic types of anaerobic fermentation, thus influencing biogas production efficiency. At present, the main metabolic paths of hydrogen production in microorganism fermentation include propionic acid fermentation, butyric acid fermentation and ethanol-type fermentation. Of them, the ethanol-type fermentation is considered as the optimal metabolic path of hydrogen production.

2.1.3.6 Pre-treatment

Pre-treatment technology is one of important factors influencing the effects of anaerobic fermentation and changes material properties through physical, chemical or biological methods. For example, macromolecular organic matters that are difficult to degrade are decomposed into micromolecular substances that are easily utilized by microorganisms, thus improving efficiency of subsequent biological treatment.

(1) Physical method

Physical pre-treatment includes the methods, such as hydrothermal hydrolysis, mechanical crushing and high-energy radiation. The hydrothermal hydrolysis pre-treatment can effectively improve biodegradability of organic wastes in villages and towns, reduce fat and salt contents in kitchen wastes, adjust nutrients and realize uniformity of physical and chemical properties. The mechanical crushing method utilizes external force to damage physical structure of substrate cellulose, while cannot eliminate lignocellulose. The high-energy radiation uses microwaves to treat substrates to improve reactivity and accessibility of materials, so as to be favorable for decomposing. This method damages internal bonds of cellulose through external force, increasing contact area of substrates between microorganisms, which is beneficial to anaerobic digestion of microorganisms

(2) Chemical method

Chemical pre-treatment is that substrates are soaked in chemical solvents, such as acid or alkali solution at a proper temperature, so that the solvents completely enter the internal structure of materials to damage reticular structure of lignocellulose. Based on this, the connecting structure of cellulose, hemicellulose and lignose is fractured, thus transforming into monose with a small molecular structure. Acid solution with a certain concentration can hydrolyse cellulose into glucose, while alkali solution with a certain concentration can promote cellulose to generate alkali cellulose. Some specific acid, alkali and salt solutions can be dissolved easily by entering crystalline structure of celluloses to generate solvent swelling. The main purpose of the chemical pre-treatment is to dissolve cellulose, hemicellulose and some lignoses through chemical reagents, thus improving the efficiency of anaerobic fermentation of substrates.

(3) Biological method

The principle of biological pre-treatment is that microorganisms or bioenzymes act on lignocellulose and their hyphae can dissolve waxy structure on the surface of plants containing cellulose and massively reproduce and grow after entering the plants. In the meanwhile, these including cellulase, hemicellulase, endo-xylanase and exoglucanase are produced, which degrades lignocellulose, thus improving utilization efficiency of substrates.

2.2 Pre-treatment Technologies of Anaerobic Fermentation

2.2.1 Hydrothermal Pre-treatment Technology of Putrescible Wastes

In view of characteristics including high contents of water, fat and salt of putrescible wastes that are unfavorable for anaerobic fermentation, the hydrothermal pre-treatment technology was used to research kitchen wastes. Through the research, the influences of various parameters of hydrothermal pre-treatment on transformation laws of substances, such as solid and liquid fat, protein and floating oil of kitchen wastes were analyzed.

2.2.1.1 Influences of Hydrothermal Pre-treatment on Crude Protein

Crude protein is one of major objective products of transforming biomasses by using the hydrothermal pre-treatment technology. Solubility of protein in water increases as ionisation of water strengthens, while ionisation of water rises with growing temperature and pressure. Therefore, raising treatment temperature and shortening treatment time are important conditions for obtaining high-yield protein (Fig. 2.2). The research showed that the content of crude protein rose after kitchen wastes underwent hydrothermal pre-treatment at 80 and 120 °C. At this time, hydrolysis of colloid of crude protein exerted no effects on higher structure of protein and hydrolysis degree is low. In addition, some macromolecular substances, such as carbohydrate and lipid in kitchen wastes were liquefied or leached higher than crude protein, so that the content of crude protein increased. When temperature rose further, the effects of hydrothermal pre-treatment on peptide bonds strengthened, which changed internal structure and significantly increased solubility. Therefore, at 150 and 200 °C, the

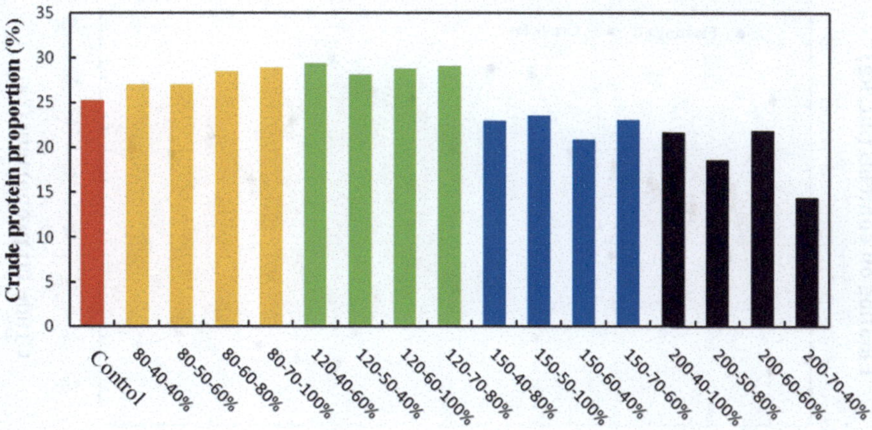

Fig. 2.2 Changes of crude protein in solid phase under different treatment conditions

content of solid crude protein relatively decreased, and kitchen wastes treated for 70 min at 200 °C after adding 40% of water had the lowest content of crude protein, which reduced by 44.97% in comparison with the original samples.

2.2.1.2 Effects of Hydrothermal Pre-treatment on Fat

Floating oil is one of important factors influencing biodegradability of kitchen wastes. A lot of long-chain fatty acids (LCFAs) in floating oil can be adsorbed on the surface of cells to restrict the transport of nutrients to cells, thus inhibiting growth and reproduction of microorganisms and influencing subsequent biodegradable efficiency of kitchen wastes. The content of floating oil in kitchen wastes undergoing hydrothermal pre-treatment significantly increased (Fig. 2.3). The output of floating oil in the experimental group treated at 120 and 150 °C was relatively higher, while that in the experimental group at 80 and 200 °C was lower. The highest output of floating oil (67.7 mL/kg) was achieved by adding 40% of water and being treated for 60 min at 150 °C, which rose 2.65 times compared with that of the control group. At this time, the content of solid crude fat reduced by 37.29%. Obviously, floating oil is closely related to the content of crude fat. Crude fat produces fatty acids and glycerin after hydrothermal hydrolysis and fatty acids can generate esterification reaction with saccharides to further promote hydrolysis of fat. In addition, diffusion between free oil and water in solid phase of kitchen wastes enhances after hydrothermal hydrolysis, which accelerates free oil to form floating oil, so that the content of solid crude fat decreases.

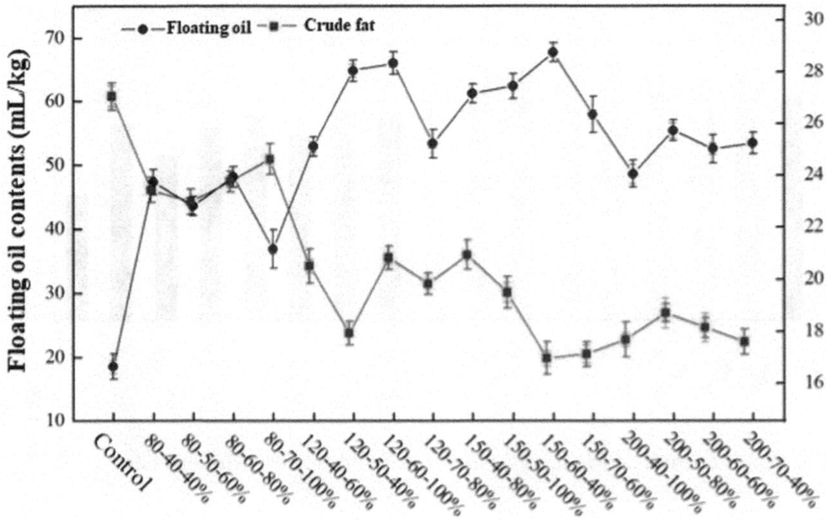

Fig. 2.3 Impacts of different hydrothermal hydrolysis pre-treatments on floating oil and crude fat

2.2.1.3 Impacts of Hydrothermal Pre-treatment on Saccharides

Saccharide, as an important carbon source for the growth of microorganisms, can be used directly or transformed into other raw materials or energy substances and is mainly produced by hydrolysis of carbohydrates. When carbohydrates collide with hydronium ion (H_3O^+) or hydroxy radical (OH^-), glycosidic bonds fracture to form the corresponding monosome, namely, reducing sugar. The research demonstrated that the content of reducing sugar in the experimental group treated at 80 and 120 °C decreased, because low temperature resulted in the low hydration and ionisation degree and material system was diluted after adding water. However, the content of reducing sugar in the experimental group treated at 150 and 200 °C rose. The content of reducing sugar added with 40% of water and treated for 60 min at 150 °C was the highest, 0.257 g/L, which increased by 16.29% in comparison with the control group (Fig. 2.4). Obviously, in hydrothermal system of hydrothermal pre-treatment, the higher the temperature was, the greater the average kinetic energy of molecular motion, the higher the hydration and ionisation degree and the stronger the effects on glycosidic bonds. When the treatment temperature was above 150 °C, the increase of liquefaction degree of carbohydrates raised the content of reducing sugar. The total sugar showed a same change trend with reducing sugar and increased by 38.92% in comparison with the control group. The total sugar mainly included reducing sugar and non-reducing sugar. In non-reducing sugar, starch, sucrose and cellulose generated saccharification reaction under the effects of thermal hydrolysis to form soluble monosaccharides and oligosaccharides. On the one hand, saccharides were important organic substrates in biotreatment. On the other hand, saccharides were important energy substances in life activities of microorganisms, while metabolic

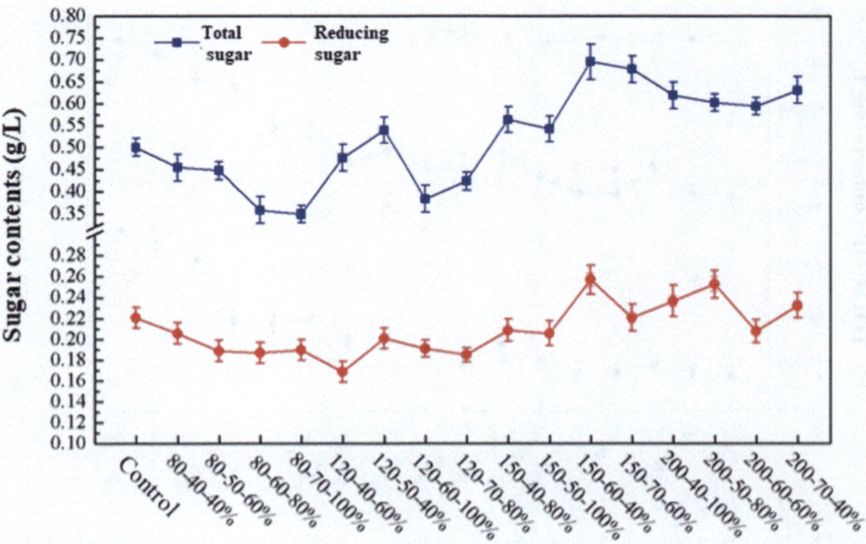

Fig. 2.4 Influences of different hydrothermal hydrolysis pre-treatments on total sugar and reducing sugar

activity of microorganisms directly affected biotreatment efficiency. Therefore, the increase of content of saccharide substance in liquid phase of kitchen wastes has important influences on the subsequent anaerobic fermentation.

2.2.1.4 Effects of Hydrothermal Pre-treatment on Organic Carbon

With increasing hydrothermal temperature and prolonging treatment time, the content of organic carbon in liquid phase constantly rose. After adding 40% of water and being treated for 70 min at 200 °C, the contents of total organic carbon (TOC) and chemical oxygen demand (COD) reached the highest of 23.7 and 109.7 g/L, which raised by 1.67 and 1.39 times in comparison with the control group without pre-treatment. Furthermore, the lowest ratio of volatile solid (VS)/total solid (TS) in solid phase was 69.04%, which decreased by 15.43% compared with the control group without pre-treatment. VS/TS is an important index to measure the transformation of solid organic matter into liquid phase of kitchen wastes. As shown in Fig. 2.5, the changes of VS/TS after hydrothermal treatment was negatively correlated with TOC and COD and as heating time prolongs and temperature rises, the less the water was added, the more significant the reduction of VS/TS. Obviously, after hydrothermal hydrolysis pre-treatment, solid substances in kitchen wastes were constantly liquefied, dissolved and transferred to the liquid phase, which provided more nutrition substrates with high homogenisation for anaerobic fermentation.

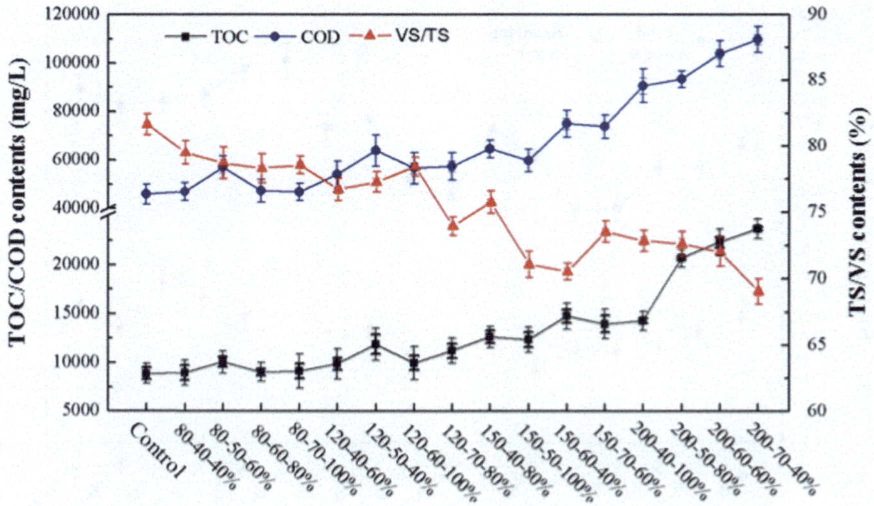

Fig. 2.5 Impacts of different hydrothermal hydrolysis pre-treatments on TOC, COD and VS/TS

2.2.1.5 Influences of Hydrothermal Pre-treatment on Dissolved Organic Matter

The transform of humus macromolecular substances in liquid-phase and protein-like during hydrothermal pre-treatment is described using DOM. Because of containing aromatic residues, numerous crude proteins have endogenous fluorescence. Figure 2.6 displays the analysis of the structural and compositional characteristics of the dissolved organic matter (DOM) during the hydrothermal pre-treatment using the excitation-emission matrix (EEM) with fluorescence regional integration (FRI). The spectra all corresponded to the existence of disparate fluorophores characterized by their excitation/emission (Ex/Em) wave-length pairs. It can be clearly identified that there were four peaks in the fluorescence spectra of EEM in the initial extracts. The first main peak (peak A), observed at the Ex/Em wave-length of 280/340 nm, represents protein-like and tryptophan substances. The second main peak (peak B) located at Ex/Em of 225/330 nm was caused by the fluorescence of tyrosine. The third broad peak (peak C) was found at Ex/Em of 360/450 nm. The final broad peak (peak D) at Ex/Em of 270/450 nm occurred because of the visible fluorescence of fulvic-like substances.

The fluorescence parameters (peak locations and fluorescence intensities included) of the spectra were summarized. Compared with the control, the location of peak A was found red-shifted to longer wavelengths (by 10–30 nm) after undergoing the hydrothermal pre-treatment. A red shift is associated with the existence of carbonyl containing substituent, hydroxyl, amino groups, alkoxyl, and carboxyl constituents. However, due to the carbohydrates and protein, Millard reactions took place and generated melanoidins slightly harder to be biodegraded. In comparison

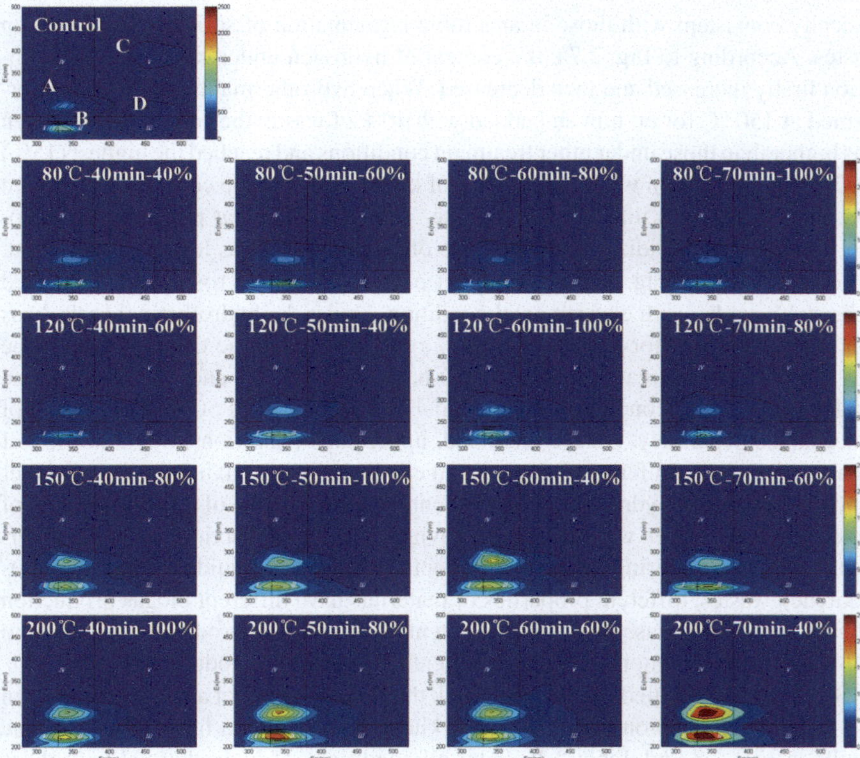

Fig. 2.6 The fluorescence EEM of the DOM fractions in different hydrothermal pre-treatments

with the control, the peak B also red-shifted to longer wavelengths (by 5–25 nm). Besides, the intensity of fluorescence peak in regions I and II was higher than that in other regions. As temperature was raised and heating intervals were prolonged, it can be seen that peak B showed a continuously increasing intensity and the highest fluorescence intensity was 2.5 times that of the control. According to the results, the hydration ionisation was enhanced substantially, which prompted liquefaction of crude proteins added with 40% of water below 200 °C for 70 min. Peaks C and D disappeared after hydrothermal pre-treatment, indicating hydrothermal pre-treatment being capable of humic acids, degrade fulvic acid, and other refractory macromolecular materials, thus strengthening their bioavailability.

2.2.1.6 Impacts of Hydrothermal Pre-treatment on Anaerobic Fermentation

The changes of hydrogen and methane contents during anaerobic fermentation of mixed phase of kitchen wastes through different hydrothermal pre-treatments are basically consistent with those in anaerobic fermentation of solid phase of kitchen wastes. According to Fig. 2.7a, the content of hydrogen under each treatment condition firstly increased and then decreased. When hydrothermal hydrolysis was performed at 150 °C for 60 min and added with 40% of waste, the content of hydrogen was higher than those under other treatment conditions and reached the highest (15%) at 72 h. In comparison with liquid phase of kitchen wastes, the content of hydrogen was about 5–6 times that of mixed phase. This indicates that potential of hydrogen production by acidizing mixed phase of kitchen wastes is low to some extent. In accordance with Fig. 2.7b, the content of methane firstly rose and then tended to be stable under each experimental condition. Under the hydrothermal hydrolysis condition at 150 °C for 60 min with 40% of water added, the content of methane was the highest and reached 58% at 110 h, followed by 55 and 54% of methane produced under the conditions at 200 and 120 °C for 70 and 50 min with 40% of water added separately. On the one hand, under such conditions, the inhibition of high content of oil on fermentation system can be reduced maximally. On the other hand, hydrothermal hydrolysis pre-treatment improves degree of homogenisation of mixed phase of kitchen wastes, thus improving methane production in anaerobic fermentation. By comparing anaerobic fermentation of solid, liquid and mixed phases of kitchen wastes, different properties of materials have different biogas production efficiencies; liquid phase shows higher potential of hydrogen production, while solid and mixed phases demonstrate higher potential of methane production.

By using the modified Gompertz model, the kinetic curves of anaerobic hydrogen and methane production of liquid-phase kitchen wastes were fitted. Based on the results in Fig. 2.8 and Table 2.1, under all conditions of hydrothermal hydrolysis pre-treatment, R^2 of the fitted curve of hydrogen production was larger than 0.99, while that of the fitted curve of methane production was larger than 0.98. After adding 40% of water and being treated at 150 °C for 60 min in hydrothermal hydrolysis, the highest hydrogen production was only 8.21 mL/g.VS and the maximum potential for hydrogen production was 23.01 mL. Under this condition, the highest cumulative

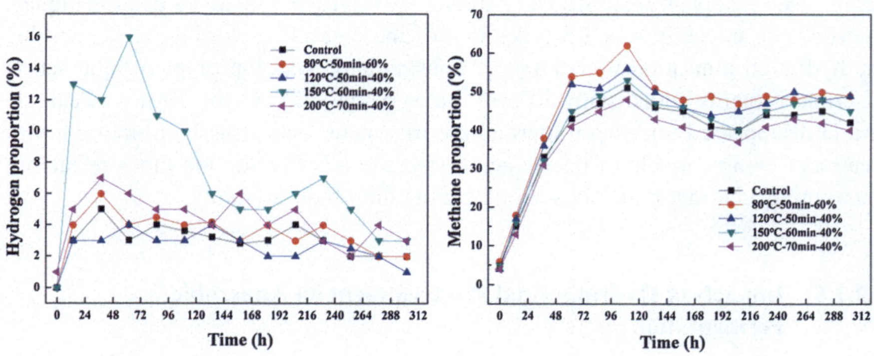

Fig. 2.7 Effects of different treatment conditions on hydrogen and methane proportion of kitchen waste mixed phase anaerobic fermentation

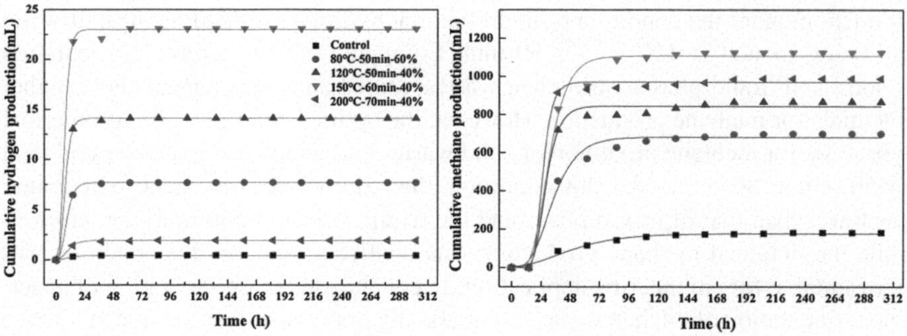

Fig. 2.8 The Gompertz model of accumulation hydrogen and methane production

Table 2.1 Kinetic analysis of anaerobic fermentation gas production

Pre-treatments	Hydrogen				Methane			
	[a]P_{max} (mL)	[b]R_{max} (mL/h)	[c]λ (h)	R^2	[a]P_{max} (mL)	[b]R_{max} (mL/h)	[c]λ (h)	R^2
Control	0.45	0.03	3.19	0.99814	178.15	12.53	9.53	0.98201
80–50–60%	1.71	0.75	3.56	0.99978	683.21	18.57	14.93	0.98463
120–50–40%	14.13	1.72	3.29	0.99963	843.48	43.43	18.02	0.99118
150–60–40%	23.01	3.03	3.45	0.99772	1108.58	69.31	16.44	0.99648
200–70–40%	1.96	0.14	3.51	0.99514	971.94	39.35	18.31	0.98881

[a]P is the biogas production potential (mL)
[b]R is the maximum biogas production rate (mL/h)
[c]λ is the lag phase time (h)

methane production was 400.17 mL/g.VS and the maximum potential for methane production was 1,108.58 mL/g.VS. In comparison with the original samples, after hydrothermal pre-treatment, the strating time for hydrogen and mehthane-producing stages both prolongs. The prolonging was not so obvious in the hydrogenogenic stage, while that was more significant in the methanogenic stage. Under the condition of hydrothermal hydrolysis at 150 °C for 60 min with 40% of water added, the starting time of methane production was 18.02 h, which was about 1.9 times that of the original samples without hydrothermal treatment. As for the highest biogas production rate, after hydrothermal hydrolysis pre-treatment, hydrogen production rate of liquid phase of kitchen wastes relatively increased, while the methane production rate improves significantly. On hydrothermal hydrolysis conditions of adding 40% of water and being treated at 150 °C for 60 min, the highest methane production rate reached 69.31 mL/h, which improved by 5.6 times in comparison with the original samples. This was followed by those under the condition of hydrothermal hydrolysis of adding 40% of water and being treated at 120 and 200 °C for 50 and 70 min, where the highest hydrogen production rates were 43.43 and 39.35 mL/h separately. Therefore, the liquid phase of kitchen wastes presents the highest potential for methane production under the condition of hydrothermal hydrolysis of adding 40% of water and being treated at 150 °C for 60 min. Compared with anaerobic fermentation of solid and liquid phases of kitchen wastes, mixed and solid phases show higher potentials for methane production. However, the optimal condition of hydrothermal hydrolysis for mehtane prodution of solid wastes was to add 60% of water and treat for 50 min at 80 °C. Under this condition, the required pre-treatment temperature was lower than that of mixed phase and the required energy consumption was low, while the obtained methane production was relatively high, so this condition was more suitable for engineering application. In conclusion, phase separation of anaerobic fermentation of kitchen wastes can markedly improve utilization rate of kitchen wastes and maximise the differentiated resourceful utilization of wastes.

2.2.2 *Pre-treatment Technology of Wastes with High Content of Lignocellulose*

Agricultural activities, fruit and vegetable wholesale markets and flower markets in villages and towns can produce a lot of organic wastes with high content of lignocelluloses. Due to the lack of efficient pre-treatment technologies, it is difficult for microorganisms to utilise these wastes, thereby leading to a low resourceful utilization rate. Through the pre-treatment technologies with acid, alkali and enzyme, this study improved bioavailability and energy conversion efficiency of wastes with high content of lignocellulose by optimizing combinations.

2.2.2.1 Influences of Pre-treatment on Microstructure of Straws

The microstructures of straws before and after pre-treatment were observed by using a scanning electron microscope (SEM) (Fig. 2.9). The results demonstrated that lignocellulose structures of straws after pre-treatment were damaged in different degrees. Although the structure was still complete after acid pre-treatment, waxy structure gradually disappeared and showed bubbles on the surface. After alkali pre-treatment, the surface presented pores and porous structures and was uneven. Moreover, the surface was deeply corroded, so that internal microfibre was completely exposed. Obviously, acid and alkali pre-treatments significantly damage lignocellulose structure of materials, which is favorable for microorganisms to utilize the materials.

Cellulase treatment damaged large area of fibre structure, resulting in serious structural damages. Due to deep corrosion, the structure was decomposed into fragments. Owing to cellulase can be specifically adsorbed on and thus decompose cellulose structure of materials, such that the contact area of microorganisms increased, which was beneficial to the interaction of materials and microorganisms and promoted the resourceful utilization of anaerobic fermentation.

Fig. 2.9 SEM images of straws after different pre-treatments and microbial community during the two-stage fermentation. **a** Non-pre-treatment; **b** 1 mol/L HCl pre-treatment; **c** 1 mol/L NaOH pre-treatment; **d** 10 mg/g TS R-10 pre-treatment

2.2.2.2 Effects of Pre-treatment on Anaerobic Fermentation

To evaluate their efficiency of fruit–vegetable wastes (FVWs), a range of pre-treatments of were carried out. Compared with the control, the three pre-treatments (with acid, alkali and enzyme) conducted on the FVWs in the hydrogenogenic stage led to higher biogas yields and hydrogen proportions (Fig. 2.10a). Observably, the cumulative biogas production of FVWs pre-treated with acid exhibited the highest efficiency of 730 mL, which was more than five times that of the control (125 mL). Through the pre-treatments with alkali and enzyme, biogas production volumes of 300 and 580 mL, respectively were obtained. The biogas produced in the hydrogenogenic stage was mainly composed of hydrogen and carbon dioxide. While using the acid pre-treatment, the results indicated that the highest hydrogen proportion was 41.2%. Pre-treatments of FVWs with reactors containing alkaline and enzyme reached hydrogen proportions of 16.4 and 38.1%, separately. After operating for 87 h during the methanogenic stage, the pH was adjusted to 7.5. In the process, it was computed that the maximum cumulative biogas was 1290 mL in the case that the FVWs were subjected to acid pre-treatment. The highest methane proportion was 89.0% after pre-treating the FVWs with alkali. Methane held proportions of 76.1 and 80.7% separately in the total biogas after pre-treatments with acid and enzyme. The evaluation of the overall performance indicated that the pre-treatment of FVWs with acid in the hydrogenogenic stage showed the highest efficiency in biogas production (730 mL, 41.2% hydrogen) on conditions that it was combined with the

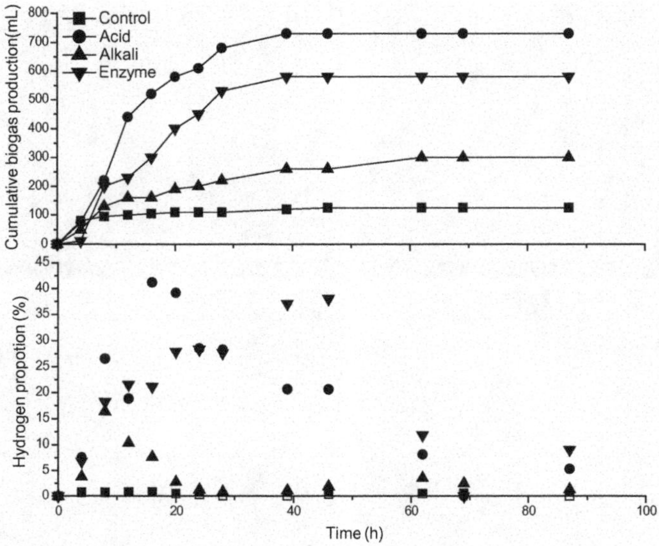

(a)

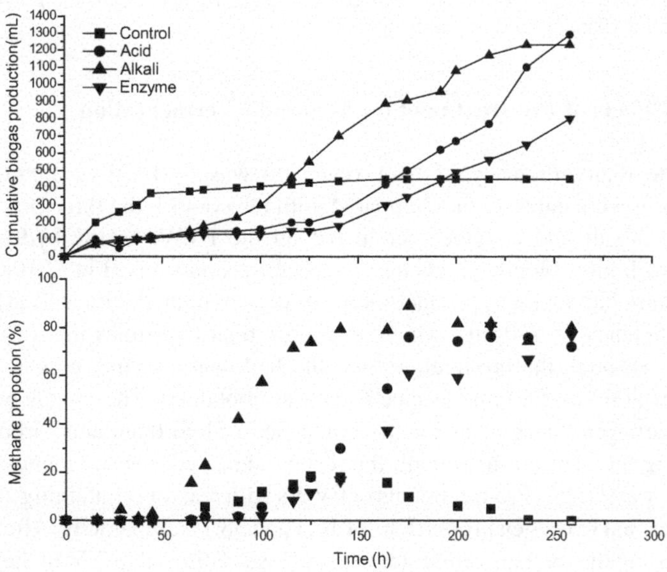

(b)

Fig. 2.10 The cumulative biogas production and proportion in response to different treatments **a** hydrogenogenic stage and **b** methanogenic stage methanogenic process (1290 mL, 76.1% methane). While after alkali pre-treatment, the volume fraction of methane was found to be precisely 89.0% in the methanogenic process during the operation containing two stages.

Based on an overall determination coefficient (R^2) of 0.96, the methane production and cumulative hydrogen on conditions applying different pre-treatments can be reflected favourably using the modified Gompertz equation for non-linear numerical estimations (Fig. 2.11). The highest potential of hydrogen production of all these three pre-treatments in the hydrogenogenic stage was higher than that of the control. The yield of hydrogen production potential of the PVWs pre-treated with acid was found to be the maximum of 180 mL, followed by 142 and 19 mL after the pre-treatments with enzyme and alkali, respectively. The maximum production rates of hydrogen of the three pre-treatment methods (with acid, alkali and enzyme) were 10.11, 3.64 and 5.65 mL/h, respectively. By contrast, the production rate of hydrogen of the control was merely 0.16 mL/h. The lag phases of the pre-treatments of PVWs with acid and alkali changed in a narrow range from 3.88 to 3.61 h and reached to the peak of 4.54 h for the pre-treatments with enzyme. In comparison, the lag phase for the control was found to be the minimum (approximately 0.38 h). In the methanogenic stage, the pre-treatments with acid and alkali were carried out similarly, during which the highest potentials of methane production were 851 and 825 mL, separately, while that in the pre-treatment with enzyme was only 332 mL. Moreover, the highest potential of methane production was lowest for the control (9 mL). The pre-treatments with acid, alkali and enzyme were separately found to have the methane production rates of 4.67, 6.63 and 2.63 mL/h. Likewise, the minimum methane production rate was observed for the control (0.15 mL/h). Meanwhile, compared with the lag phase of the control (83.75 h), those for the acid and enzyme pre-treatments were even longer (146.65 and 147.39 h, respectively). Whereas, the lag phase can be shortened on conditions that bacteria were enriched by utilizing and pre-adapting to inimical growth conditions after pre-treatment with alkali in the methanogenic stage. In this way, it can be seen that the lag time of 87.53 h during the alkali pre-treatment was similar to that of the control.

2.2.2.3 Impacts of Pre-treatment on TOC and Total Nitrogen

In the CHMP-AF of mixed wastes produced in pedlars' markets, the changes laws of COD, TOC and total nitrogen (TN) concentrations in different stages are shown in the Table 2.2. The research displayed that COD and TOC concentrations of FVWs increased in the hydrogenogenic stage, mainly because hydrolytic acidification transformed solid materials into soluble ones. In the methanogenic stage, microorganisms effectively utilized carbon sources, resulting in decrease of COD and TOC concentrations and a low utilization rate of TN in CHMP-AF. This is the main reason for easy accumulation of ammonia nitrogen and inhibition of biogas production of microorganisms.

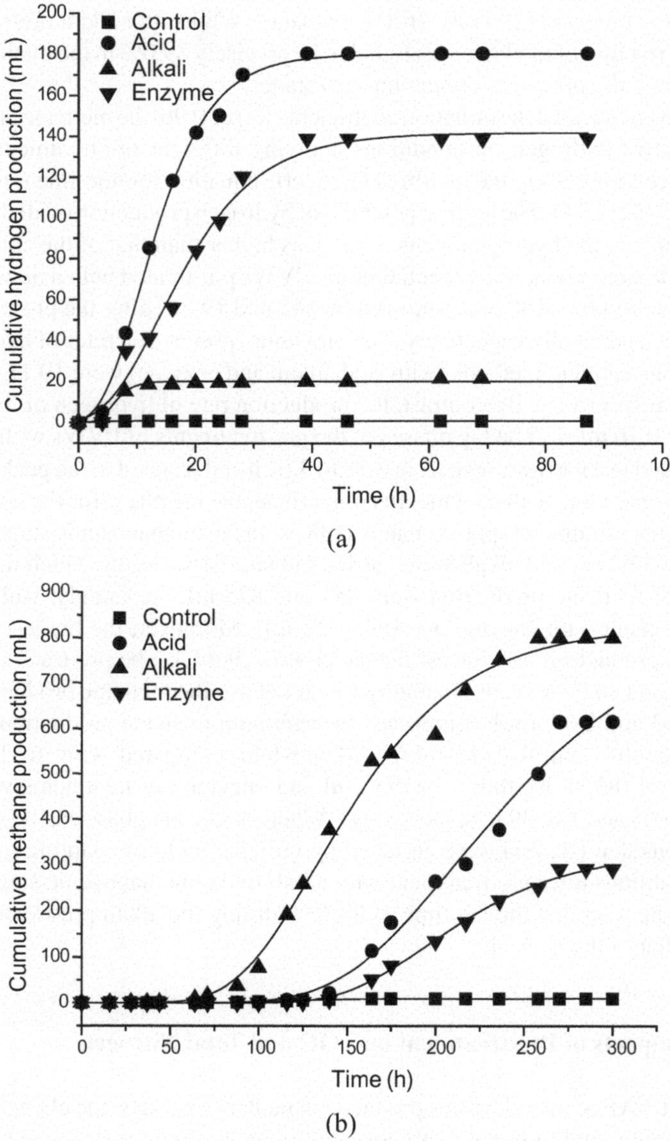

Fig. 2.11 The cumulative hydrogen and methane production in response to different pre-treatments as a function of time **a** hydrogenogenic stage and **b** methanogenic stage

2.2.2.4 Effects of Pre-treatment on Ethanol and VFA

Acetic, pentanoic, butyric, propionic, lactic, hexanoic acids, and ethanol were observed in all experiments (Fig. 2.12). Trace amounts of ethanol were tested in the control and the samples pre-treated with acid. Propionic and acetic acid were

Table 2.2 Changes of metabolites including COD, TOC and TN

Pre-treatments	COD (mg/L)			TOC (mg/L)			TN (mg/L)		
	Initial	HS	MS	Initial	HS	MS	Initial	HS	MS
Blank	7408	12,470	3030	1970	2295	1383	1130	937.1	526.4
0.25% HCL	5706	13,780	7238	2724	8624	2468	90.8	384.4	556
1% HCL	7562	12,154	10,440	2728	6492	4624	91.2	532	396
0.5% NaOH	6920	14,940	5416	2912	9444	2260	110.8	351.2	580
1% NaOH	7562	13,976	10,260	3172	8652	2404	142.4	308.8	492
3% NaOH	11,522	12,978	8920	6372	11,040	3516	696	536	612
10 mg enzyme	11,310	17,286	5700	6776	8532	3060	492	656	596

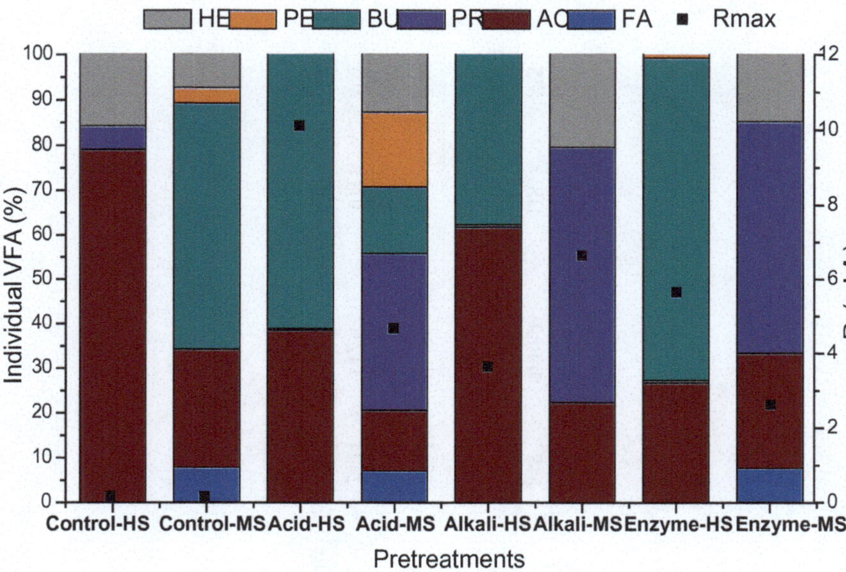

Fig. 2.12 Percentage of individual VFA accounting for the total VFAs and R_m in response to different pre-treatments with the combined hydrogen/methane production process

ranked the first in the hydrogenogenic stage according to the analysis on the VFA compositions in the two stages, while propionic acid was the leading product after the methanogenic stage. Most of the soluble metabolites of the pre-treatments with alkali and enzyme comprised butyrate and acetate, as evaluated by GC (which in combination held 99.4 and 98.7% separately of the overall soluble metabolites in the two pretreatments). The result indicated that a fermentation based on butyric acid happened in the hydrogenogenic stage. For the pre-treatments with acid, the soluble metabolites consisted of acetate, ethanol and butyrate, which totally held a proportion of 99.6% of the overall soluble metabolites. This composition implied that bothe fermentation processes based on ethanol and butyric acid took place in the hydrogenogenic systems. Nevertheless for the pre-treatments with acid, fermentation using butyric acid was the primary metabolic pathway. Moreover, it was observed that the control test led to the highest concentrations of lactic acid (3.84 g/L) in comparison with those in all other pre-treatments after the hydrogenogenic stage. This probably threatened the efficiency of anaerobic digestion after the hydrogen production. The soluble metabolites, particularly butyrate and acetate acids displayed dramatically reducing concentrations in the methanogenic stage, while the percentage of propionic acid was higher than that of all pre-treatments in the hydrogenogenic stage.

2.2.2.5 Influences of Pre-treatment on DOM

By using the blind parallel factor (PARAFAC) model, the researchers identified three separate fluorescent components from the DOM (Fig. 2.13). Component 1 is centered at 220/340 nm, the highest Ex/Em wavelength pair. In addition, it is associated with typical protein-like substances, including tryptophan explored previously in effluents of the fermentation. There are latest evidences providing support for the opinion of component 1 being probably derived from phenolic compounds having Ex/Em wavelength pairs resembling to those of component 1 and originating from the lignin degradation. Component 1 occupying a dominant F_{max} value in the DOM in the hydrogenogenic stage on conditions of the three pre-treatments is the representative protein-like substances. Nevertheless, no component 1 is found in the original FVWs (i.e. the control). This result indicates that the protein-like substances in component 1 mainly stemmed from the newly formed DOM because of microbial and biological activities which are related to the pre-treatments methods.

Component 2 comprises two excitation maxima at 225 and 275 nm, with one emission peak at 330 nm. The two components exhibit Ex/Em characteristics like those of protein-like fluorescent compounds and presumably a combination of fluorophores carrying fluorescent proteins tryptophan and tyrosine: free amino acid or

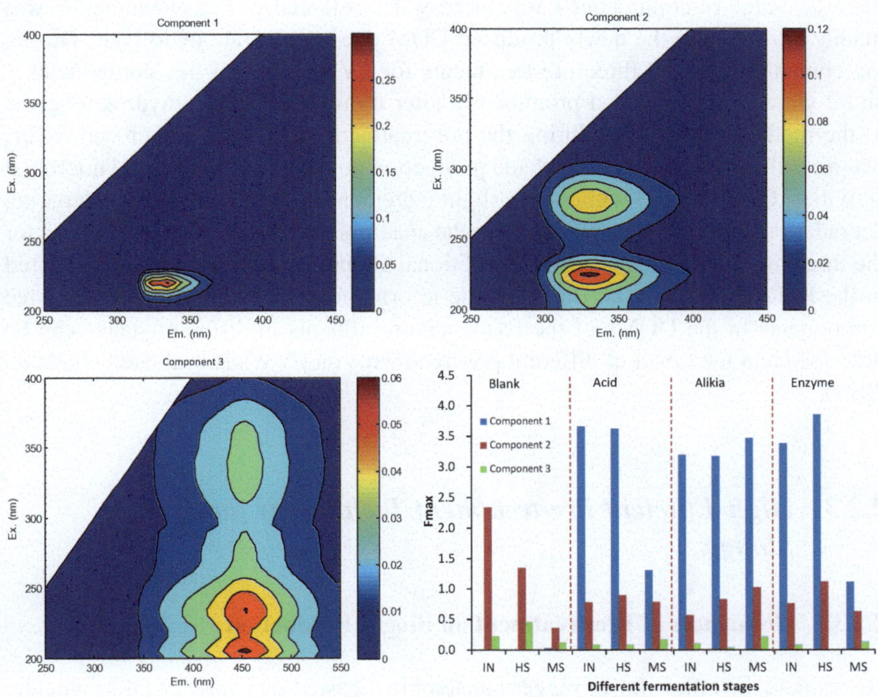

Fig. 2.13 The EEM spectra of the three component identified by PARAFAC analysis and maximum fluorescence intensities (F_{max}) in different fermentation stages (Jia et al. 2014)

amino acid that is bound in structures of protein molecules. This result probably reveals that both protein-like fluorescent components present the spectral properties of tryptophan and tyrosine bound into greater structures of organic molecules but not the pure compounds in the fermentation process of two stages. The F_{max} value of component 2 decreased during the biodegradation of FVWs in the production process of two stages except for pre-treatment with alkali. Whereas, component 2 showed a lower F_{max} value than component 1.

For component 3, it presents three fluorescence peaks, whose maximum Ex/Em wavelength pairs are centred at 205/425, 230/425 and 320/425 nm. These three peaks are humic-like peaks of short wavelengths and deriving from fulvic-like substances. The presence of the component in the three fermentation stages was related to allochthonous DOM stemming from terrestrial origins. The fulvic-like acid substance showed the lowest relative concentration in the component 3 among these three components and except for control, a slightly higher concentration was detected after methane production than that in other stages.

It can be observed from Fig. 2.13 that the F_{max} value of humic-like substances of component 3 was the lowest in all samples, whereas the highest values were found for protein-like substances of components 1 and 2. Furthermore, except for that in the control test, the abundance of component 1 was higher than components 2 and 3, which agrees well with the F_{max} values. After pre-treating the FVWs, the F_{max} value of component 1 was increased substantially. The phenomenon was mainly attributed to the newly produced DOM due to substrate hydrolysis. During the comparison of the three pre-treatments for FVWs, protein-like components 1 and 2 were found degraded prominently after transiting from the hydrogenogenic to the methanogenic stage during the pre-treatments with acid and enzyme. This was a result of the hydrogen/methane production potentials and increased microbial activities. Component 3 witnessed a slight increment in its amount after undergoing degradation and metabolic processes in the anaerobic fermentation of two stages for the three pre-treatments. Generally, additional quantitative information was offered in the PARAFAC analysis. Based on the information, the distribution of the three components in the DOMs of the fermentation effluents in different stages can be described and the effect of different pre-treatments on FVWs is evaluated (Jia et al. 2014).

2.2.3 *High-Efficient Pre-treatment Technology for Seed Sludge*

2.2.3.1 Influences of Pre-treatment on Biogas Production

To evaluate the efficiency of piggery anaerobic digested residues (PADRs) inocula, a wide variety of treatment methods were utilized. Figure 2.14 displays the effects of the different treatments on the cumulative production of biogas in batch tests of

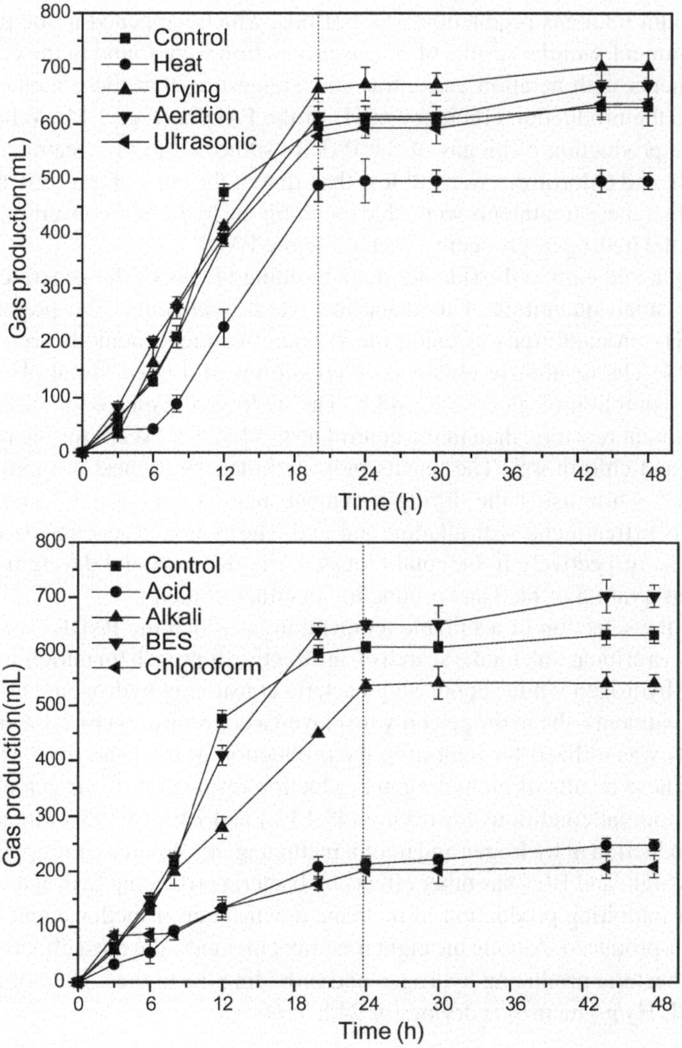

Fig. 2.14 The 48 h cumulative biogas production of the PADRs undergoing different treatments. The test conditions were as follows: batch culture, 200 mL glucose culture medium for 48 h, and an initial pH of 6.0. The data are the average of triplicates with standard deviations ($n = 3$)

48 h. Among the eight treatment methods, the physical (heat, aeration, drying, or ultrasonic) treatments resulted in superior efficiencies in cumulative biogas production to those of the chemical (alkali, acid, sodium 2-bromoethanesulfonate (BES), or chloroform) treatments. Obviously, when applying the drying and BES treatments to the pre-treated PADRs, biogas yields higher than the control can be obtained. The increment in cumulative total yield of gas was because the organic matter was transferred from particulate to soluble one. Through calculation, it is obtained that

the maximum total gas production was 704 mL, which happened in the process of drying treatment. Similar results of biogas production were found in the control test and treatments with aeration and ultrasonic, suggesting that these methods failed to enhance the production efficiency of H_2 of the PADRs (Fig. 2.14). Whereas, the cumulative production of biogas of the PADRs subjected to pre-treatment of acid, heat, alkali, and chloroform were all less than that of the control test (631 mL). This indicated that these treatments were able to inhibit the hydrogen-consuming bacteria as well as the hydrogen-producing bacteria in the PADRs.

Hydrogen and carbon dioxide are main building blocks of the generated biogas, as well as small quantities of methane in several treatments. The proportions of H_2 and CH_4 on conditions of using the various treatment methods are illustrated in Fig. 2.14. The results are obtained on conditions of a fixed initial pH (6.0) and 10 g/L consumption of glucose for 48 h. The hydrogen displayed higher fractions in the treatment reactors, than in the control test (32.42%), except for the treatments with acid and chloroform. The results indicated that the highest proportion of H_2 was 49.13% while using the drying treatment method, and 47.23% and 31.88%, respectively in treatments with alkaline and acid. The hydrogen accounted for 48.20% and 49.13%, respectively in the total biogas during the heat and drying treatments, without observation of methane production in either of them.

Hence, the selection of a suitable temperature at which the PADRs were treated thermally is a critical link for the search of an effective approach for enriching bacteria producing hydrogen while suppressing bacteria consuming hydrogen. Dissimilar to the BES treatments, the hydrogen only displayed a concentration of 2.1% in the case that $CHCl_3$ was utilized for inhibiting the production of methane from acetate and H_2/CO_2. These results of biohydrogen production revealed that one can effectively apply the optimal conditions for drying (49.13%) and BES (46.93%) treatments to prompt production of hydrogen and inhibit methanogens. Whereas, the application of BES costs high, and BES has other effects on bacteria producing hydrogen: it is also capable of inhibiting production of methane downstream if methane and hydrogen were the co-produced. Among the eight treatment methods, the most effective one for enriching bacteria producing hydrogen and providing the highest yield of hydrogen (172.94 mL H_2/g glucose) is drying for 24 h at 60 °C.

2.2.3.2 Effects of Pre-treatment on pH

It can be seen that the final pH changed in the range of 4.11–5.68 for each treatment reactor. Among the eight treatments, the drying, BES, and aeration treatment methods showed the final pH values of 5.68, 5.67, 5.66, respectively, which were approximate to 5.59, final pH of the untreated PADRs. In comparison, for fermentation reactors using the PADRs treated with acid, alkaline, and heat the final pH values were all below 5.0, which inhibited the production of bacteria producing hydrogen. The optimal pH value for hydrogen production is in the range of 5.5–6.0. These results indicated that, due to anaerobic fermentation, the various treatments led to dissimilar buffering capacity in the production systems of bio-hydrogen.

2.2.3.3 Impacts of Pre-treatment on Ethanol and VFAs

The soluble metabolites were investigated at the end of each batch (Table 2.3), which revealed that the primary VFAs and ethanol showing varying distributions for different pre-treatments. The primary soluble metabolites generated in the drying and alkali pre-treatments included butyrate and acetate, as determined by GC (accounting for 85.2 and 91.9% separately of the overall soluble metabolites generated in the two reactors). The result suggested that butyric acid fermentation was formed in the two reactors. The soluble metabolites produced in the pre-treatment cases with heat and acid included ethanol, acetate, and butyrate, altogether accounting for 99.5 and 97.5% separately of the total soluble metabolites generated in the two reactors. It suggested the occurrence of both ethanol and butyric acid fermentation processes in these fermentation systems. The compositions of VFA treated with $CHCl_3$ resemble to those in initial tests, that is, $CHCl_3$ inhibited the synthesis of acetate from CO_2 by cells of *Clostridium thermoaceticum* during the glucose fermentation. In the control and the pre-treatment based on aeration, lactic acid was found to be the dominant soluble metabolite, separately taking up 71.17 and 80% of the end products of the liquid fermentation, implying a typical fermentation process of ethanol. It can be seen from these results that the various pre-treatment methods probably pertained to the diverse microbial populations detected in the inocula after enrichment.

In the reactors for drying, acid, alkali, and heat pre-treatment, no hydrogen was consumed to alleviate the feedback inhibition of VFA production. Therefore, much higher amounts of acetic acid and ethanol were produced than those in the PADRs pre-treated with ultrasonic waves, aeration, BES, and chloroform, as well as the untreated one. However, an intermediate homoacetogenesis process might exist during the production of methane from hydrogen and VFAs from the PADRs pre-treated with aeration, ultrasonic wave, and the control. Moreover, it is interesting that among all pre-treatment methods, the pre-treatment with aeration led the lactic acid to have the

Table 2.3 The soluble metabolite concentrations produced after the different treatments

Pre-treatments	Volatile fatty acids concentration (mg/L)						
	ET	AC	PR	BU	PE	HE	LA
Initial	6.58	413.43	83.00	531.44	44.18	20.26	0.00
Control	299.52	790.61	23.74	394.06	278.86	157.52	4799.60
Aeration	314.46	998.77	18.79	415.13	180.69	99.10	5370.67
Ultrasonic	249.41	684.29	80.45	1357.66	124.72	95.40	563.43
Heat	782.80	978.59	4.62	1137.94	6.50	2.60	0.00
Drying	348.84	1117.34	131.12	1838.88	3.20	28.83	0.00
Acid	903.80	1316.16	3.68	957.46	53.13	24.44	0.00
Alkali	234.19	1546.56	58.58	1814.54	1.99	2.52	0.00
BES	271.89	722.49	86.32	1193.71	111.27	75.29	807.78
Chloroform	87.99	644.24	6.14	361.26	69.73	46.92	70.94

highest concentration (5.37 g/L), probably threatening the efficiency of anaerobic digestion.

2.2.3.4 Influences of Pre-treatment on DOM

Figure 2.15 illustrates that the 3D fluorescence spectra of EEM can distinguish the fermentative hydrogen production in the eight pre-treatments. Whereas, the intensity of fluorescence spectra and ranges of peak wavelengths exhibited remarkable differences: it is observed that under all pre-treatments, the fluorescence peaks in Regions I and IV were higher than those in Regions II, III, and V. This revealed that the produced metabolites mainly involved biodegradable DOM and slightly non-biodegradable DOM. The former included tyrosine-like and by-product-like materials and the latter consisted of tryptophan-like, fulvic acid-like, and humic acid-like substances. Or, metabolites were not produced in the process of anaerobic fermentative hydrogen production. In comparison with the control, the pre-treatments with

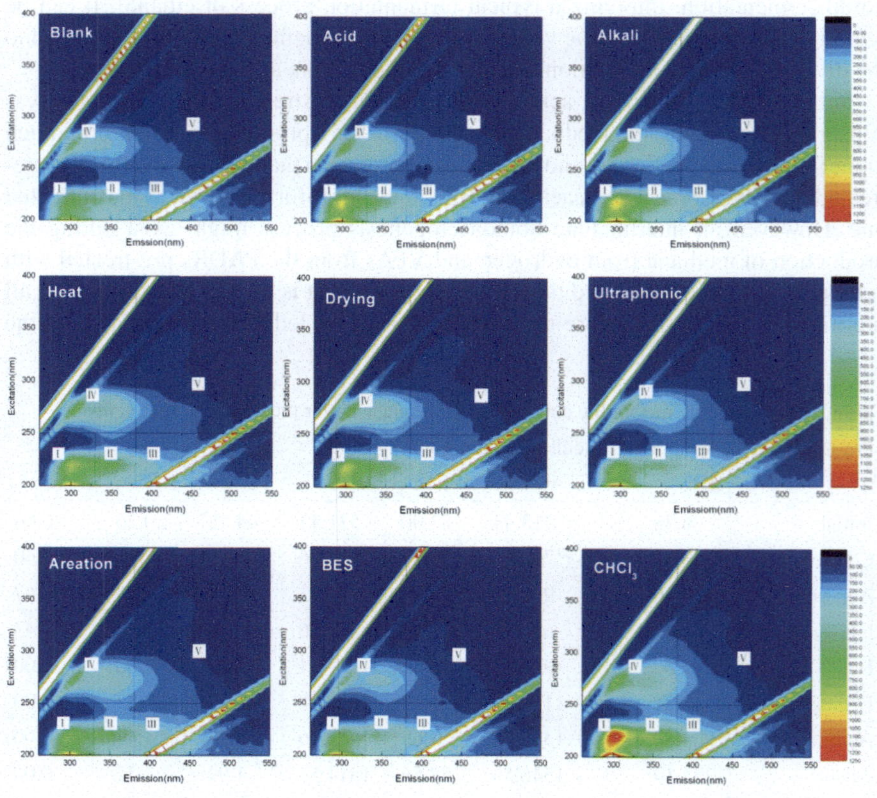

Fig. 2.15 The fluorescence EEM of the DOM fractions that were extracted from the differently pre-treated PADRs (Jia et al. 2013)

acid, alkali, and $CHCl_3$ displayed the most intensive protein-like peaks at the Ex/Em wavelengths of 220/300 nm. In both Regions I and II, the protein-like peaks exhibited lower intensity and number in the other pre-treatment methods. Particularly in the drying pre-treatment, the peaks of the by-product-like materials and tyrosine-like were centred on the Ex/Em wavelength pairs of 220/305, 270/300 and 270/305 nm. Whereas, no humic acid-like matter peak was observed.

It can be learnt from the above results the control bioreactor was found to show complicated microbial community, in addition to diverse fluorescence peaks. In the sludge treatments with acid, alkali, and $CHCl_3$, microorganic activity was suppressed, resulting in incomplete metabolism, strong peaks of biodegradable protein-like materials, and a lower characteristic DOM peak in the 3D fluorescence spectrum. The bacteria consuming hydrogen were inhibited in the BES and drying pre-treatments which while improved activity of bacteria producing hydrogen, giving rise to adequate substrate metabolism. Hence, the drying of the PADRs led to the rapid and complete consumption of the biodegradable DOM, along with the lowest content of organic material hard to be degraded. Therefore, drying is the optimal pre-treatment method for PADRs. Additionally, it is demonstrated in the research on the EEM spectra of DOM that the substrate metabolism generated some biodegradable protein-like materials and lower non-biodegradable materials. These substances themselves were potential for methane production through anaerobic digestion (Jia et al. 2013).

2.3 Research on High-Efficient Anaerobic Fermentation Technology for Organic Wastes

2.3.1 High-Efficient Anaerobic Fermentation Technology of Kitchen Wastes

Based on hydrothermal pre-treatment technology of kitchen wastes, the coupling technologies and equipment of high-efficient hydrothermal pre-treatment and anaerobic fermentation of kitchen wastes were developed (Fig. 2.16). This study deeply discussed the influences of different conditions of hydrothermal pre-treatment on transformation characteristics of organic wastes and the performance of combined hydrogen and methane production. Furthermore, this research elaborated the change laws of metabolic intermediates in combined production via anaerobic fermentation, succession of microbial community, and response relation of functional protein and analysed metabolic mechanism of combined hydrogen and methane production of organic wastes. Based on the functional regulation level of protein, the optimisation and regulation strategies of the process for improving hydrogen production efficiency and stable operation were proposed. This solves the difficulties, such as long starting time, low biogas production efficiency and unstable operation in the actual operation.

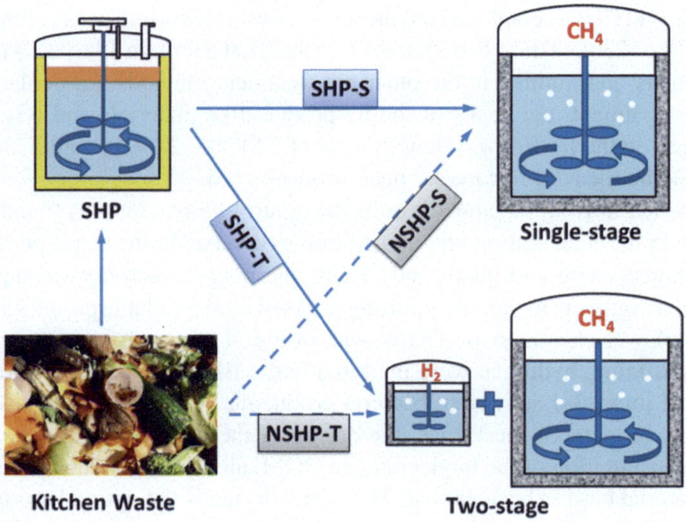

Fig. 2.16 Experiment design and schematic diagram

2.3.1.1 Performance of Biogas Production

To explore the effects of SHP on AD processes of single and two stages, four processes were designed, including SHP-S, NSHP-S, SHP-T, and NSHP-T. They separately represent SHP and non-SHP with a single-stage AD process, and SHP and non-SHP with a two-stage AD process.

It can be seen from the performance of the biogasification process that the highest cumulative yield of biogas was 6.43 L in SHP-T, two times higher than that in the NSHP-S (i.e., 3.21 L) (Fig. 2.17a). As for the biogas productivity, it was 6.7 mL/ (L h) in SHP-T. The highest methane proportion was found in SHP-T to be 76.75% and approximate values of 74.70 and 71.66% were separately obtained in NSHP-S and NSHP-T (Fig. 2.17b).

It is reported in Table 2.4 that the process of two stages was greater than the single-stage process in terms of the highest biogasification potential and rate. The highest production rates of biohydrogen and methane were 13.33 and 15.81 mL/h separately in SHP-T (Fig. 2.17c, d). The treatment using the SHP displayed a shorter time of lag phase compared with the control in the methanogenic stage, a rate-limiting step of biogasification. This suggested that a short-term hydrothermal pre-treatment was able to dramatically enhance the biodegradability and homogenisation of FW, leading to a quick initiation of methane production and stable utilization of substrate.

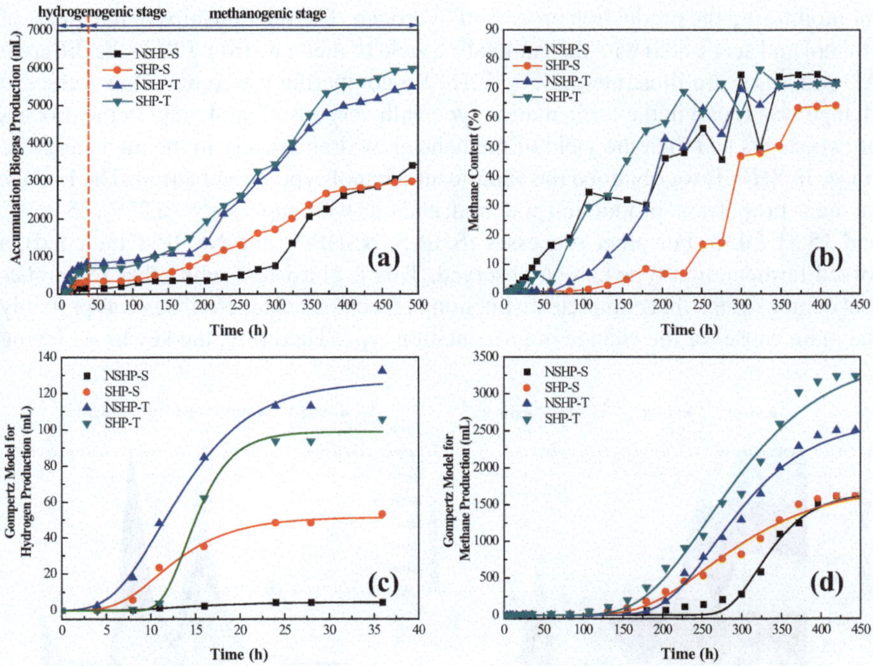

Fig. 2.17 Biogasification performance and kinetic analysis during the different AD processes. **a** Accumulative biogas production; **b** methane proportion; **c** accumulative biohydrogen production; **d** accumulative methane production

Table 2.4 Kinetic analysis of biohydrogen and methane production in the different AD processes

AD processes	Hydrogenogenic stage			Methanogenic stage		
	[a]P_{max} (mL)	[b]R_{max} (mL/h)	[c]λ (h)	[a]P_{max} (mL)	[b]R_{max} (mL/h)	[c]λ (h)
SHP-S	51.53	4.01	6.01	1759.62	8.03	152.50
NSHP-S	4.36	0.30	7.50	1658.25	10.32	282.54
SHP-T	98.88	13.33	11.25	3519.38	15.81	175.47
NSHP-T	126.59	8.39	5.60	2638.16	15.07	202.17

[a]P is the biogas production potential (mL)
[b]R is the maximum biogas production rate (mL/h)
[c]λ is the lag phase time (h)

2.3.1.2 Ethanol and VFA Contents

During the anaerobic digestion processes, the generation of VFAs and ethanol typically accompanied the formation of hydrogen. Therefore, the concentrations and compositions of the generated soluble metabolites can serve as effective indicators for monitoring the production process of hydrogen. The relationships of the yields of ethanol and acetic acid with the cumulative yield of methane from FW in the different AD processes are illustrated in Fig. 2.18. By comparing different AD processes, a change was found in the fermentation type with varying ethanol yield in the process of two stages and then the yield of methane grew dramatically in the methanogenic stage. In SHP-T, we obtained the significant ethanol-type fermentation. The highest methane proportion, production potential, and rate were separately 76.75%, 3519.38, and 15.81 mL/h. For other processes (SHP-S, NSHP-S, and NSHP-T included), a mixed fermentation type can be observed. This is consistent with other researches that demonstrated that complete distinction of feedstock characteristics was probably the main cause for the change in fermentation type. Therefore, the key to achieving

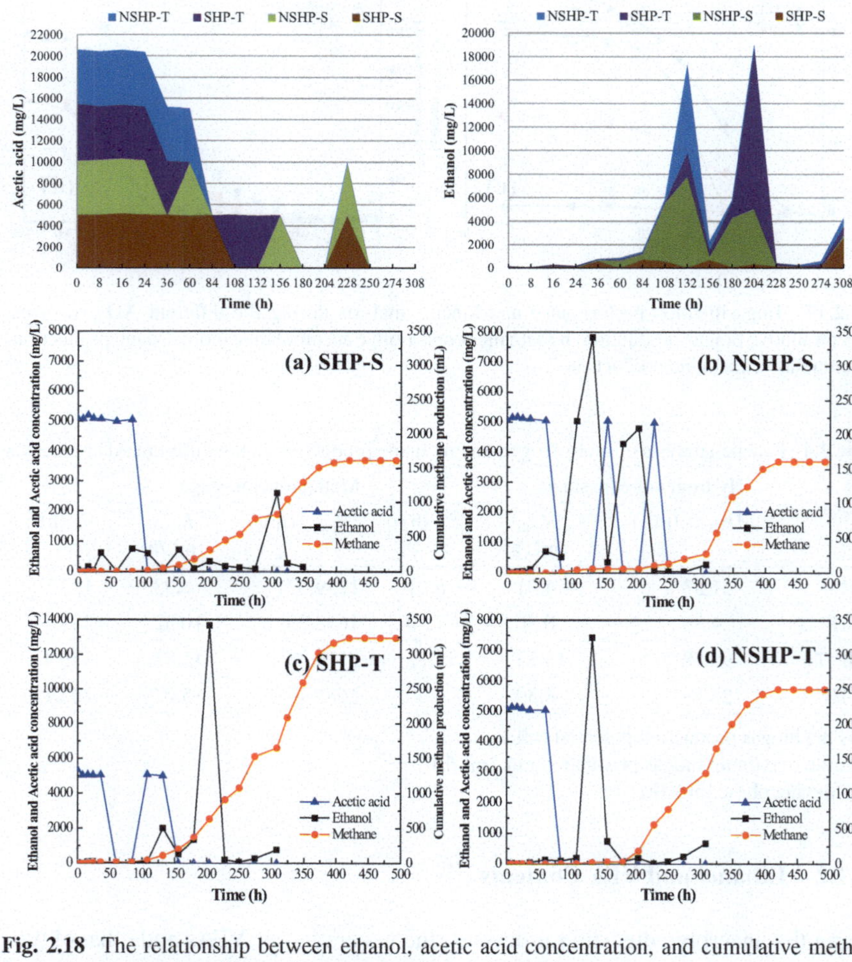

Fig. 2.18 The relationship between ethanol, acetic acid concentration, and cumulative methane production during the different AD processes (Jia et al. 2017a)

good performance in the methanogenic stage during a two-stage process is to control the fermentation type in the hydrogenogenic stage by SHP.

2.3.1.3 DOM

The DOM in the effluent of fermentation obtained from the anaerobic fermentation process of two stages was composed of lots of fluorescence compounds, and fluorescence peaks of different compounds were probably overlapped. PARAFAC analysis is a multivariate chemometric method capable of decomposing EEM spectra into various fluorescent components independent of each other. It offers the accurate quantitative information on the component distribution in the DOM fractions. The researchers identified three components, marked 1, 2, and 3, from the EEM spectra by the FEMS-PFA and in Fig. 2.19 the fluorescence intensity is expressed as F_{max}. Component 1 consisting of two fluorescence peaks is located at the highest

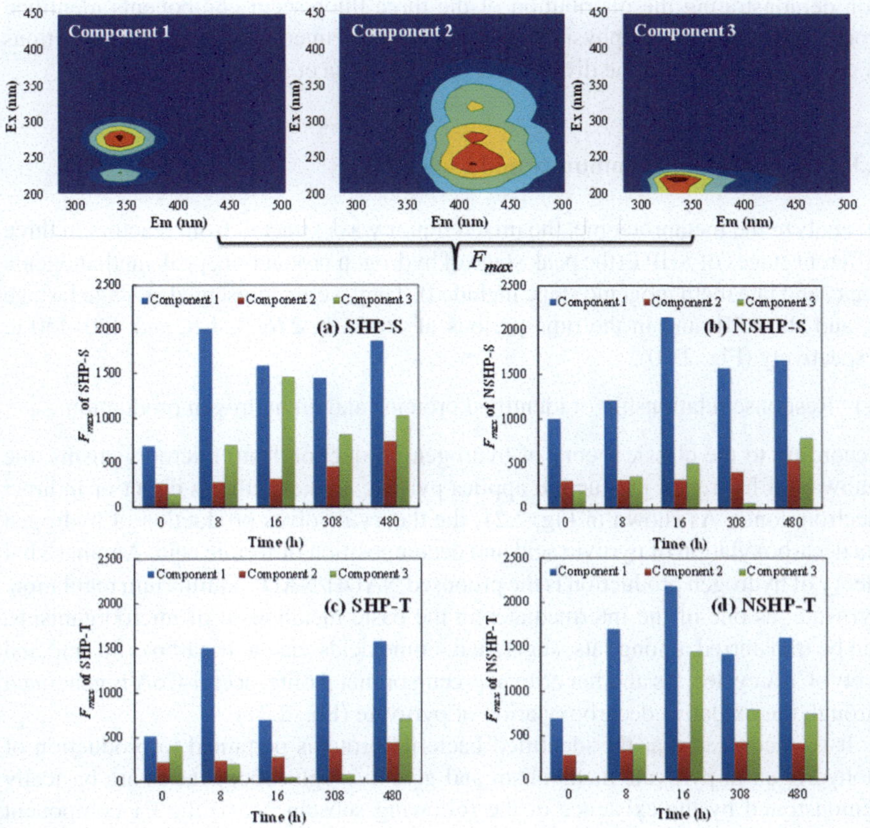

Fig. 2.19 F_{max} of the three fluorescent components by the FEMS-PFA from the different AD processes (Jia et al. 2017a)

E_x/E_m wavelength pair of 225/340 and 280/340 nm. It was a combination of tyrosine protein-like and tryptophan materials according to the result. The main peak at 280/340 nm and identified in Component 1 refers to the attributes of soluble microbial by-product-like material. For Component 2, the highest E_x/E_m wavelength pairs were found at 240/420, 280/420, and 320/420 nm in the three fluorescence peaks. The component was reported to be basically characterized as humic-like matter. Component 3 was observed at the highest E_x/E_m wavelength pair at 220/330 nm. Regarded to derive from tryptophan protein-like materials, it was characterized by the short-wavelength absorption.

It can be seen from the fluorescence intensity of F_{max} that the tryptophan and tyrosine protein-like materials the dominated the fluorescence spectra of DOM. Through the comparison of different AD processes, the component 2 in SHP-T and NSHP-T exhibited lower F_{max} values than those in the processes of a single stage. Due to the increasing hydrolysis of the biorefractory DOMs and microbial activities under SHP, protein-like component 1, together with component 3, was degraded significantly via SHP-T. Basically, the FEMS-PFA results provided additional quantitative information demonstrating the distribution of the three fluorescent components identified from the DOMs and the physiological responses of microbes to varying conditions in their environment in the diverse AD processes (Jia et al. 2017a).

2.3.1.4 Microbial Community and Function

To analyze the metaproteome, the mixed liquor was extracted from reactors in three different stages of SHPT (the peak stage of hydrogen production, peak methanogenic stage, and late methanogenic stage included). They were represented as stage I, stage II, and stage III, and in the time periods of 16–24 h, 276–324 h, and 420–450 h, respectively (Fig. 2.20).

(1) Response relationship of identified proteins and biohydrogen production

According to the classic theory of hydrogen production from microorganisms, the pathway of hydrogen production applies pyruvic acid as either a direct or indirect electron donor. As shown in Fig. 2.21, the theory involves production of hydrogen via decarboxylation of pyruvic acid and decomposition of formic acid. Another vital theory of hydrogen production is the proposed $NADH/NAD^+$ equilibrium regulation. Pyruvate, as one of the intermediates in the basic metabolism of microorganisms, can be transferred among fats, sugar, and amino acids via the tricarboxylic acid and acetyl-CoA cycles. As another critical precursor metabolite, acetyl-CoA is generated through the oxidative decarboxylation of pyruvate (Fig. 2.21).

It is uncovered that the identified bacterial proteins pertained to production of biohydrogen in pyruvate metabolism and glycolysis/gluconeogenesis are basically demonstrated by the existence of the following substances: (i) the E1 component

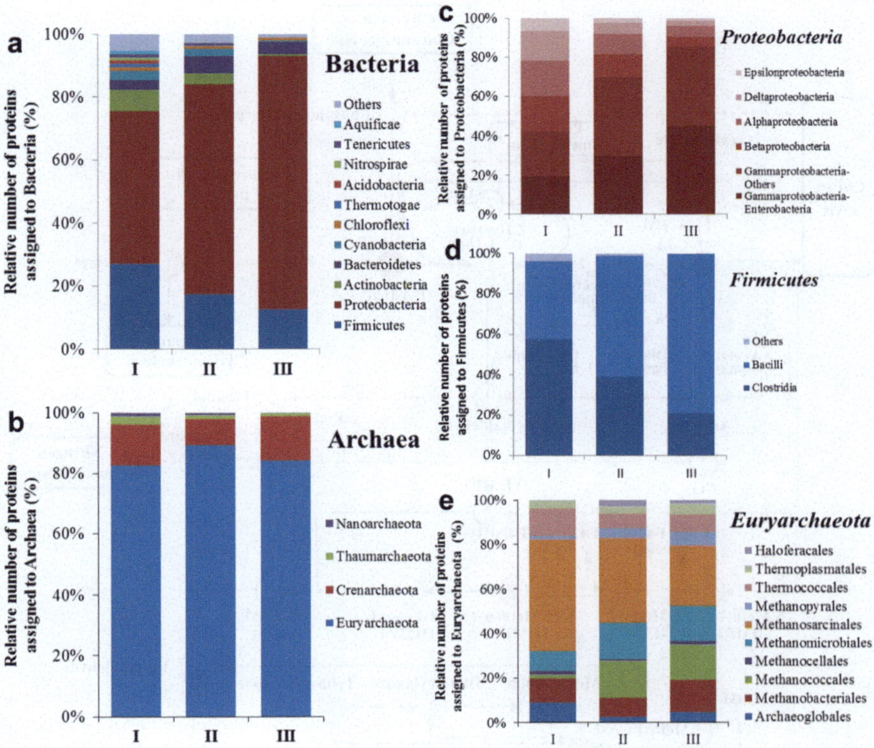

Fig. 2.20 Relative number and distribution of bacterial (**a**) and archaeal (**b**) proteins in different stages of the SHPT process, based on taxonomic bins for proteome-derived proteins with taxonomic annotation. Distributions in **c** and **d** are at the level of classes within the phyla Proteobacteria and Firmicutes, respectively. **e** Is at the level of the orders within the Euryarchaeota (Jia et al. 2017b)

of pyruvate dehydrogenase; (ii) dihydrolipoamide dehydrogenase; (iii) pyruvate-ferredoxin/flavodoxin oxidoreductase; (iv) formate C-acetyltransferase; and (v) acetyl-CoA synthetase, as displayed in Fig. 2.22. The E1 component of pyruvate dehydrogenase found in Stage II and involved in pyruvate metabolism, is able to change pyruvic acid into acetyl-CoA and CO_2 and stems from *Escherichia*. The dihydrolipoamide dehydrogenase appears in the decarboxylase pathway of pyruvic acid for biohydrogen production belonging to *Pseudomonas* and *Zymomonas* in Stages II and III, separately. Pyruvate-ferredoxin/flavodoxin oxidoreductase which is found in *Pelosinus* and *Kluyvera* bacteria in Stage I is a critical enzyme for the equivilibrium regulation of NADH/NAD^+ in hydrogen production. These results indicated that although these identified proteins belonged to disparate genera, they probably can happen simultaneously with biogydrogen production together. This pathway contains non-native hydrogenase linked with NAD(P) and H_2 generation from NAD(P)H provided rapidly from glucose degradation via the pentose phosphate pathway. Formate C-acetyltransferase (pyruvate formate lyase, PFL) belongs

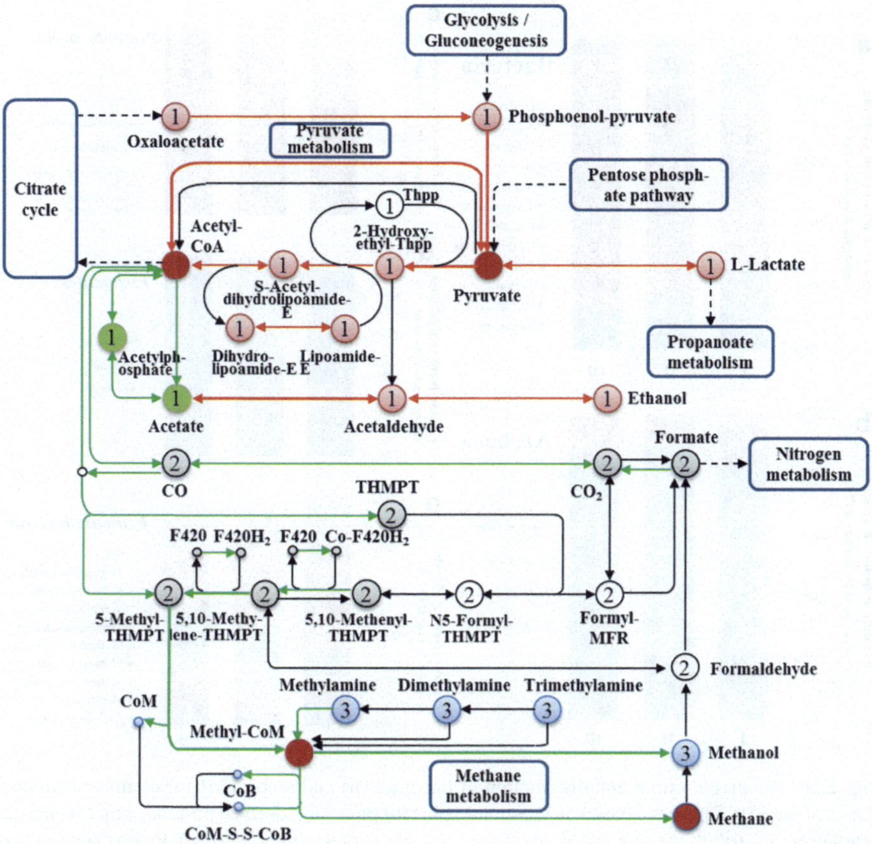

Fig. 2.21 Proposed metabolic pathways of biohydrogen and methane coproduction from FW during SHPT as inferred from the metaproteomic data (Jia et al. 2017b)

to *Staphylococcus* and *Escherichia* in Stage II and assists the regulation of anaerobic glucose metabolism. Based on the radical non-redox chemistry, the substance catalyzes the reversible transformation of CoA and pyruvate into acetyl-CoA and formate. Afterwards, formate dehydrogenase and hydrogenase show catalyzation effect on the oxidation of formate to CO_2 and hydrogen, and donate electrons to a second substrate, representing another vital pathway for hydrogen production in pyruvate metabolism. The results indicated that the protein of major subunit of formate dehydrogenase was affiliated to *Methanotorris* in *Euryarchaeota* in archaea and *Escherichia* in *Gammaproteobacteria* in bacteria identified in Stages I and II, separately. There was also research concluded that *Escherichia coli* produced hydrogen from formate based on the formate hydrogen lyase complex composing formate dehydrogenase-H and hydrogenase 3.

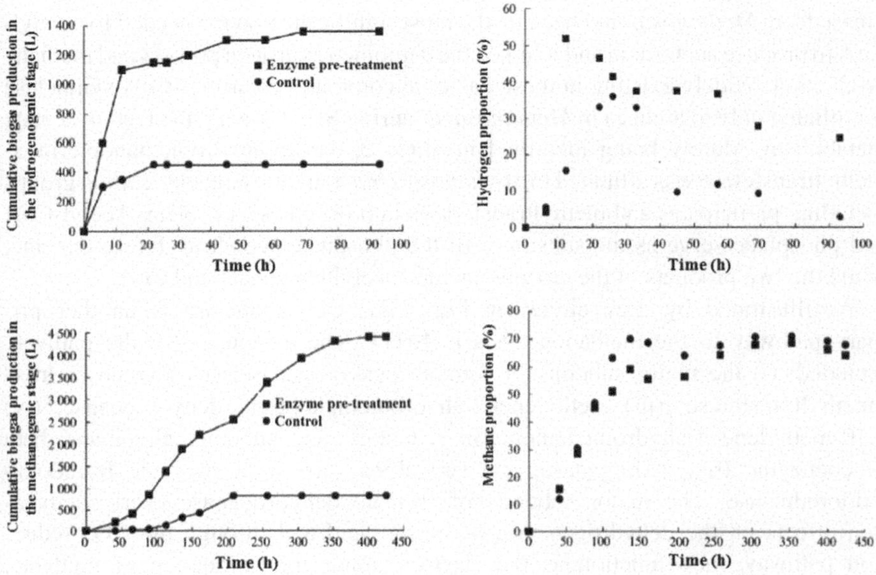

Fig. 2.22 Changes of cumulative biogas production and biological gas production in different stages of CHMP-AF of straws

(2) Response relationship of identified proteins and methane production

Production of methane from microorganisms is a process capable of reducing methyl in compounds containing one carbon to methane under joint actions of many coenzymes and enzymes in one-carbon metabolism. This was realized by the cooperation of plenty of archaea proteins and small quantities of bacterial proteins. This study revealed that during SHPT, three pathways, i.e. acetate, CO_2 reduction, and methylotrophic pathways, separately involving 10, 6, and one protein for methanogenesis were active. In addition, it can be seen that the proteins in the acetate pathway held the vast majority of the total activity. Observably, expression of acetyl-CoA decarbonylase/synthase complex (sub-units alpha, beta, delta, epsilon, and gamma) was found at higher levels in SHPT, especially in Stage II, during which acetic acid was used as the solo carbon source and energy to catalyze the decarboxylation of acetyl-CoA as it was decomposed to CO_2. Therefore, it is a critical enzyme for the pathway through which methanogenesis uses acetate.

Other proteins mainly involved in the acetate pathway included: (i) sub-units A and H of tetrahydromethanopterin S-methyltransferase; (ii) phosphate acetyltransferase; (iii) acetate kinase; and (iv) acetyl-CoA synthetase. Remarkably, based on the metaproteomic analysis, these enzymes all belonged to Euryarchaeota and to *Methanosarcina*, and *Methanosaeta*. It was found in Stages II and III that acetate kinase from *Methanosarcina* became the most significant enzyme needed for acetyl-CoA to produce acetic acid and it linked the combined hydrogen process and methane production. Widely existing in these bacterial communities, acetyl-CoA synthetase is affiliated only to archaea in *Methanosaeta* during SHPT, being involved in acetate metabolism. Merely being identified in Stage 2, the tetrahydromethanopterin S-methyltransferase was affiliated to *Methanosarcina*, transforming one-carbon groups including participates and methyltransferases in biosynthesis of folate. Acetyl-CoA and phosphate serve as the substrates of the phosphate acetyltransferase enzyme, while the two products of the enzyme include acetyl phosphate and CoA.

As illustrated by grey circles in Fig. 2.21, CO_2 reduction is another primary pathway to the methanogenesis in SHTP. The evidences for the pathway included: (i) the major subunits of formate dehydrogenase; (ii) glycine hydroxymethyltransferase; (iii) methylenetetrahydromethanopterin dehydrogenase; (iv) 5,10-methylenetetrahydromethanopterin reductase; (v) subunits alpha and beta of coenzyme F_{420} hydrogenase; and (vi) alpha subunit of pyruvate ferredoxin oxidoreductase. The major subunits of formate dehydrogenase exert catalyzation effects on the dehydrogenation of formic acid and through the CO_2 reduction pathway, they function as the electron donor for production of methane. Glycine hydroxymethyltransferase appeared in *Leclercia* and *Acinetobacter* bacteria in Stages II and III. Methylenetetrahydromethanopterin dehydrogenase exerted effects on the CH–NH group of donors with other acceptors and belongs to *Methanosarcina* in all of the three stages. The enzyme has the two substrates of 5,10-methylenetetrahydromethanopterin and coenzyme F_{420}, whereas 5,10-methenyltetrahydromethanopterin and reduced coenzyme F_{420} are the two products. The interesting thing is that 5,10-methylenetetrahydromethanopterin reductase and alpha subunit of pyruvate ferredoxin oxidoreductase were merely obtained in Stage II and belonged to *Methanothermobacter* in archaea. The identified proteins sub-units alpha and beta of coenzyme F_{420} hydrogenase were separately affiliated to *Methanothermobacter and Methanococcus*, which functioned as a deazaflavin analogue, and used formic acid and hydrogen as electron donors to generate methane by reducing CO_2.

According to the blue circles in Fig. 2.21, a primary methylotrophic pathway was revealed by the identified proteins of methylamine–corrinoid protein Co-methyltransferase assigned to *Methanosarcina*. This protein produces methane through reduction of methyl with hydrogen in methyl compounds or via the dismutation of methyl compounds. Moreover, three proteins of subunit of methyl-Co M reductase were identified and found belonging to *Methanothermobacter, Methanosarcina,* and *Methanotorris* in phylum *Euryarchaeota* during SHPT. It is the terminal methyl carrier and the critical enzyme for all three pathways of methane production (Jia et al. 2017b).

2.3.2 High-Efficient Anaerobic Fermentation Technology of Straws

In view of characteristics of materials containing high content of lignocellulose and barriers of biological metabolism in anaerobic fermentation, the combined pre-treatment and resourceful utilization technologies of anaerobic fermentation were researched and integrated. This study deeply researched the key influencing factors of fermentation technologies for the coupling system of hydrogenogic and methanogenic phases. By taking the goal of maximizing resource utilization of green wastes with high content of lignocellulose and minimizing the cost, this research established a system of resourceful utilization technologies of wastes containing high content of lignocellulose in villages and towns, which provides a basis for industrial application.

2.3.2.1 Biogas Production Performance

The CHMP-AF was performed on reed straws after being treated with cellulose R-10 and the results are shown in Fig. 2.24. In the early hydrogenogenic stage, biogas production rapidly rose in the group treated with enzyme and then slowly increased after 12 h. The maximum cumulative biogas production in the hydrogenogenic stage was 1360 mL, which was three times that of the control group without pre-treatment. Volume fraction of hydrogen in the treatment group with enzyme reached the highest of 52.1% after 12 h and then gradually decreased with reaction. In the methanogenic stage, cumulative biogas production in the group treated with enzyme stably rose with time and reached the highest, being 4400 mL after 353 h, which was five times that of the control group. Volume fraction of methane significantly increased with time. The maximum volume fraction of methane in the control group was 68.4%. Obviously, cellulase can be specifically adsorbed on cellulose structure of straws, promoting the dissolution of plant cell walls. This can dissolve intracellular substances out and degrade macromolecular polysaccharide, protein and lipid into micromolecular substances, thus improving utilization efficiency and potential of biogas production of substrates. After pre-treating straws by using cellulase R-10, hydrogen and methane productions markedly increased.

2.3.3 TOC and TN Contents

Table 2.5 shows the utilization rates of soluble chemical oxygen demand (sCOD), TOC, TN and reducing sugar in reed straws in anaerobic fermentation treated with enzymes. The results demonstrated that utilization rates of sCOD, TOC and TN in the methanogenic stage of anaerobic fermentation were significantly higher than those in the hydrogenogenic stage. Moreover, utilization rates of all indictors after

Table 2.5 Utilization rates of sCOD, TOC, TN and reducing sugar in fermentation metabolites of straws

Pre-treatment	Utilization rate in hydrogenogenic stage %				Utilization rate in methanogenic stage %			
	Reducing sugar	TN	TOC	sCOD	Reducing sugar	TN	TOC	sCOD
L-B	31.66	14.82	7.1	37.99	33.88	38.62	40.35	90.37
L-E	48.88	36.87	32.82	64.68	53.50	44.58	55.76	98.71

pre-treatment were higher than those in the control group. Utilization rate of sCOD of reeds in the hydrogenogenic stage was 64.68% and then reached 98.71% in the methanogenic stage. Owing to hydrogenogens and methanogens consume sCOD in substrates to realize cell proliferation and generate biological gases in the CHMP-AF, the utilization rate of sCOD is directly related to microbial activity. The higher the utilization rate of sCOD is, the more obvious the microbial activity and the more significant the biogas production effects of anaerobic fermentation. The experimental results present that after pretreatment with enzymes, metabolic activity of microorganisms in fermentation tanks in the hydrogenogenic stage significantly increased. However, as large amounts of volatile acids were produced in the fermentation tanks, pH of the system rapidly reduced, such that some microorganisms did not adapt and died, so removal rate of sCOD slightly decreased in comparison with that in the hydrogenogenic stage. The utilization rate of TN of reed straws in the fermentation was higher than that in the control group and reached 44.58%. TOC was the total amount of organic carbon in substrates. After treating the reed straws with 10 mg/g of R-10, utilization rate of TOC was larger than that in the control group and reached 55.76%. In addition, the utilization rate of reducing sugar in the reed straws increased by 19.7% in comparison with that without treatment and reached 53.5%, which was far higher than that in the control group.

The results of kinetic analysis on CHMP-AF demonstrated that the anaerobic fermentation of aquatic plants via the combined production of hydrogen and methane after pre-treatment with cellulase can effectively avoid inhibition of hydrogen partial pressure on methanogens. Moreover, this can significantly improve methanogenic efficiency and obtain the optimum cumulative methane production, methane content and the highest utilization efficiency of reducing sugar, sCOD and TOC. Therefore, this ensures metabolic balance of microorganisms in the CHMP-AF and effectively solves the problems, such as low biogas production efficiency and instable biogas production process of anaerobic fermentation.

2.3.3.1 Ethanol and VFA Contents

In the fermentation of straws pre-treated with enzymes, acetic acid appeared in metabolites at 65 h in the methanogenic stage and then disappeared after maintaining for 70 h. Ethanol began to increase at the end of the hydrogenogenic stage and in the beginning of the methanogenic stage. At 89 h in the methanogenic stage, ethanol content reached the highest, being 3079.7 mM and then rapidly decreased. The contents of other organic acids showed no significant changes in the reaction. The results showed that the products of anaerobic fermentation of straws pre-treated with enzymes mainly include acetic acid and ethanol, which belonged to a typical ethanol-type fermentation path. The main fermentation products included ethanol, acetic acid, CO_2 and H_2. The changes of production of volatile acids were mainly because microorganisms in substrates produced VFAs through hydrolysis and acidification of insoluble particulate organic matter at the beginning of fermentation; after that, in the late stage of fermentation, methanogens transformed acetic acids and hydrogen into methane and carbon dioxide, so that the concentration of organic acids decreased.

The content of ethanol in metabolites of the control group reached 3952.7 mM at 12 h and decreased when hydrogenogenic stage ended. In the methanogenic stage from 65 to 137 h, the content of ethanol suddenly rose and then rapidly reduced and finally decreased to the lowest after 305 h. The content of acetic acids slightly increased at 34 h in the hydrogenogenic stage and then quickly dropped to the lowest. Furthermore, butyric acid slightly rose from 65 to 113 h in the methanogenic stage, while other volatile acids did no change significantly. According to the main products of fermentation, the fermentation process of reeds in the control group mainly belonged to ethanol-type fermentation, accompanying with butyric acid-type fermentation (Fig. 2.23).

2.3.3.2 DOM

As displayed in Fig. 2.24, by using PARAFAC model, three fluorescent components are identified in the samples and their values of fluorescence intensity (F_{max}) in different stages of anaerobic fermentation are demonstrated in Fig. 2.24. After the pre-treatment, the values of fluorescence intensity gradually decreased with anaerobic fermentation. The results showed that the reason for the decrease of fluorescence intensity was that macromolecular substances in pre-treated plants were easily decomposed by microorganisms.

Component 1 shows excitation wavelength (λ_{Ex}) at 220 and 270 nm and emission wavelength (λ_{Em}) at 325 nm and mainly comprises fluorescence of protein, so it belongs to tyrosine-like fluorescent protein. Owing to protein came from microorganisms in the early stage of fermentation, the component was mainly generated from metabolic substrates or decay process of microorganisms. In accordance with F_{max}, protein-like substances represented by component 1 were the main component in all samples during anaerobic fermentation. With changing activities of microorganisms, such as hydrogenogens and methanogens in anaerobic fermentation, fluorescence intensity of component 1 gradually reduced from the highest value of 10,501.01 mM to 638 mM. The results demonstrated that in metabolites of anaerobic fermentation after pre-treatment, protein-like substances were mainly from newly formed DOM produced by biological or microbial activities.

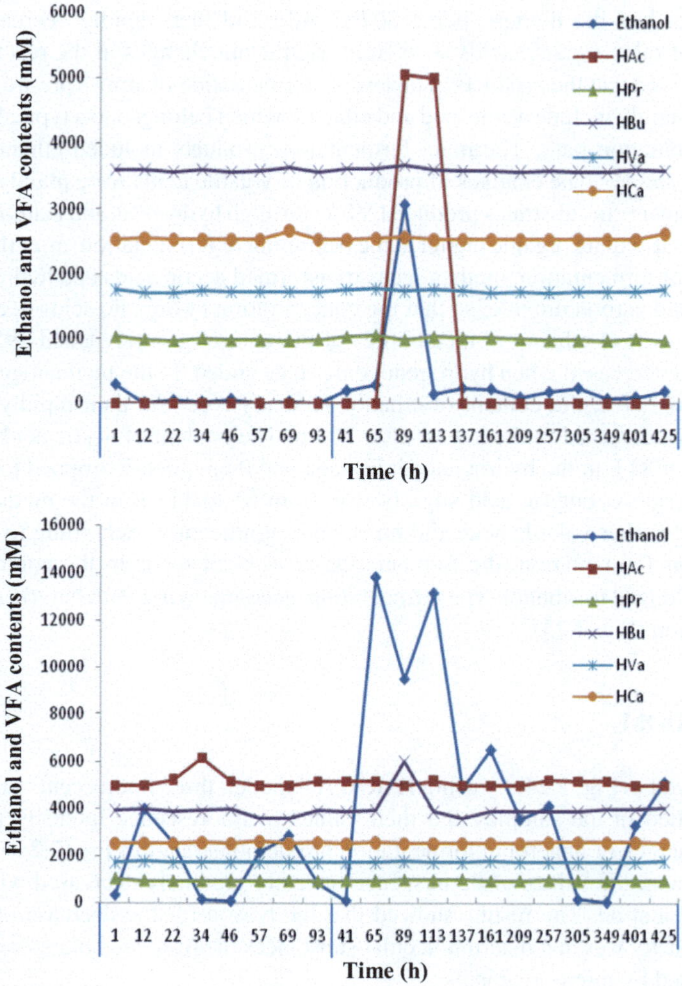

Fig. 2.23 Changes of content of volatile acids in the fermentation of straws

Component 2 has excitation wavelength at 210 nm and 270 nm and emission wavelength at 350 nm and belongs to tryptophan-like fluorescence. In the anaerobic fermentation process, the value of fluorescence intensity of protein components slightly rose. At 22 h, the value of fluorescence intensity of proteins was up to 9587.9 mM. This indicated that the reaction system was in acid environment and most of methanogens died due to low pH, so that intracellular proteins were dissolved out. Therefore, fluorescence intensity of proteins firstly increased and then decreased and finally reached 676.9 mM.

As for component 3, the excitation wavelength and emission wavelength are found at 210 and 400 nm separately and it is the typical coenzyme NADH substance. The value of fluorescence intensity also slightly increased at 22 h to 2565.4 mM and then

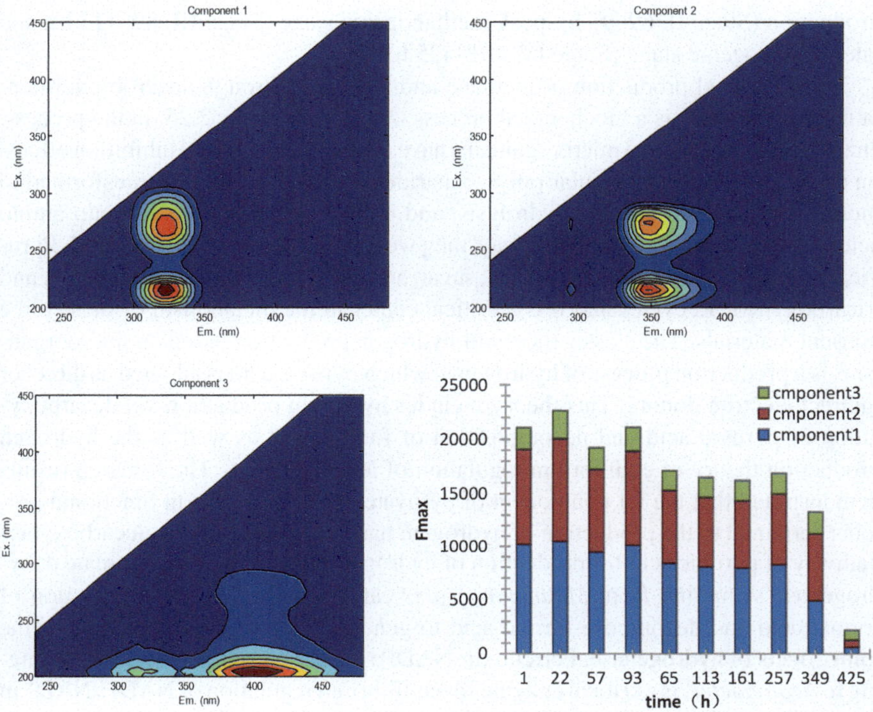

Fig. 2.24 PARAFAC analysis and F_{max} organic matter in metabolites containing three components in anaerobic fermentation (Jia et al. 2017c)

reduced to 916.1 mM. The decay of methanogens dissolved intracellular coenzyme NADH out, so that the fluorescence intensity of NADH suddenly rose. In the three components, the concentration of coenzyme NADH substance was the lowest.

PARAFAC provides additional quantitative information for describing the allocation of organic matter extracted from the samples in each stage of anaerobic fermentation. Therefore, it was observed that the changes of microbial activities of the generated protein-like substances significantly reduced. The fluorescence intensity of component 2 obviously increased and then decreased in anaerobic fermentation based on microorganisms. The accumulation of component 3 resulted in the augmentation of the fluorescence intensity of coenzyme NADH in soluble metabolites of anaerobic fermentation (Jia et al. 2017c).

2.3.3.3 Microbial Community and Function

The mixed liquor for analyzing metaproteome was collected from bioreactors in four stages, i.e. peak stage of hydrogen production (Stage I, 6–12 h), late stage of hydrogen production (Stage II, 69–93 h), peak methanogenic stage (Stage III, 89–113 h), and late methanogenic stage (Stage IV, 401–425 h).

The combined production of methane and hydrogen through anaerobic fermentation (CHMP-AF) is a biochemical process. As shown in Fig. 2.25, in the process, enzymes, substrates, and microorganisms have interaction with, and inhibition effects on each other. Macromolecular polysaccharides in reed straws were transformed to monosaccharide through the hydrolysis, and proteins were hydrolyzed into amino acids. Then via the glycolytic pathway, they were transformed to pyruvic acid. Pyruvic acid can be transferred among fats, sugar, and amino acids by the acetyl-CoA and tricarboxylic acid cycles and it is a critical center in the metabolism of these three nutrient materials. The classic theory of hydrogen production based on microorganisms is a production process of hydrogen in which pyruvic acid is adopted as direct or indirect electron donors. This theory includes hydrogen production via decarboxylation of pyruvic acid and decomposition of formic acid as well as the hydrogen production theory of equilibrium regulation of $NADH/NAD^+$. The research results demonstrated that the E1 component of pyruvate dehydrogenase, in functional proteins pertained to the production of hydrogen, has involvement in the decarboxylase pathway of pyruvic acid for production of hydrogen. Subunit alpha of formate dehydrogenase stemming from *Methanococcales* can catalyze the dehydrogenation of formic acid and decompose formic acid to generate CO_2 and hydrogen under the joint effects of hydrogenase. Ferredoxin-NADP reductase, found in bacteria belonging to *Azotobacter*, is a critical enzyme for equilibrium regulation of $NADH/NAD^+$ in the production of hydrogen. NADH is able to couple with the fermentation processes of butyric acid, propionic acid, ethanol, or lactic acid and then be oxidized as NAD^+

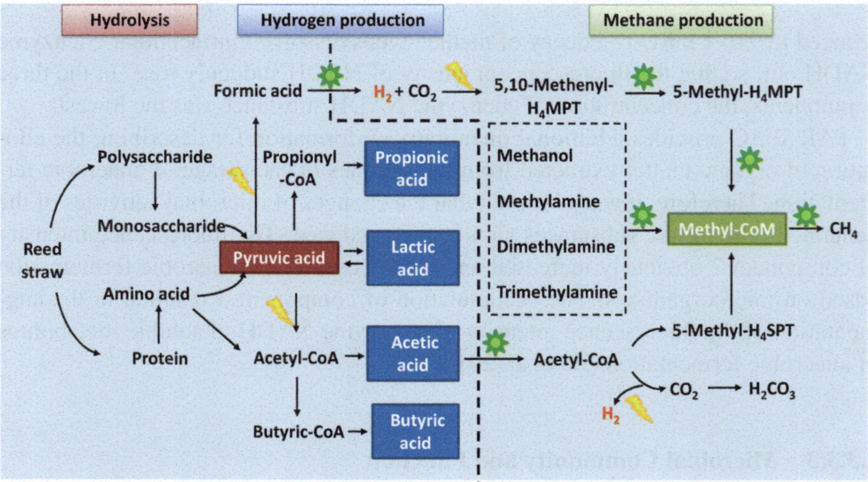

Fig. 2.25 Depiction of the metabolic characteristics of functional proteins and metabolites in different stages inferred from the metaproteome. The yellow lightning is the identified protein from bacterial community. The green sun is the identified protein from archaea community (Jia et al. 2017d)

to generate hydrogen. This ensures equilibrium of $NADH/NAD^+$ in the metabolism. Hence, many VFAs were observed in the later production stages of hydrogen and then subjected to decomposition into one-carbon compounds and CO_2 by virtue of microorganisms to be reduced for producing methane.

Production of methane from microorganisms is a process capable of reducing methyl in one-carbon compounds and CO_2 to methane under the joint action of numerous coenzymes and enzymes in the one-carbon metabolism. Coenzymes which are closely related to production of methane can be classified into two types: one is the one-carbon carriers, including H_4MPT, H_4SPT and coenzyme M; and the other is carriers of electron consisting of coenzyme F_{420}, ferredoxin, coenzyme B and cytochrome. The result indicates that the three pathways for production of methane exist in the CHMP-AF using reed straw. This was realized by the cooperation of numerous archaea proteins and small amounts of bacterial proteins, with diversity and microorganic differences in different stages.

The functional proteins relevant to production of methane via the acetic acid pathway showed the maximum proportion of the total (Fig. 2.26). Acetyl-CoA decarboxylase applies acetate as its sole source of carbon and for the energy needed for catalyzing the decarboxylation of acetyl CoA to decompose it to CO_2. Carboxyl is oxidized to produce H_2 donating electrons for the methane generation and methyl reduction. This pathway is basically realized by *Methanosarcina*, together with some archaea of *Methanosaeta*. The methanogenic stage witnessed the active methyl nutrient pathway. By using methanol, dimethylamine, methylamine, and trimethylamine as the substrates, the research identified numerous functional proteins. Due to the effect of methyl-coenzyme M reductase of the methyl carrier, generation of methane happened at the close of the production stage of methane via that pathway based on the involvement of *Methanosarcina*. By using subunit alpha of formate dehydrogenase, formic acid was subjected to decomposition to hydrogen and CO_2. Then, CO_2 was reduced under the catalysis by virtue of ferredoxin-NADP reductase in bacterial protein to produce 5,10-methenyl-H_4MPT, and then 5-methyl-H_4MPT because of the catalysis of 5, 10-methylenetetrahydromethanopterin reductase of those existing

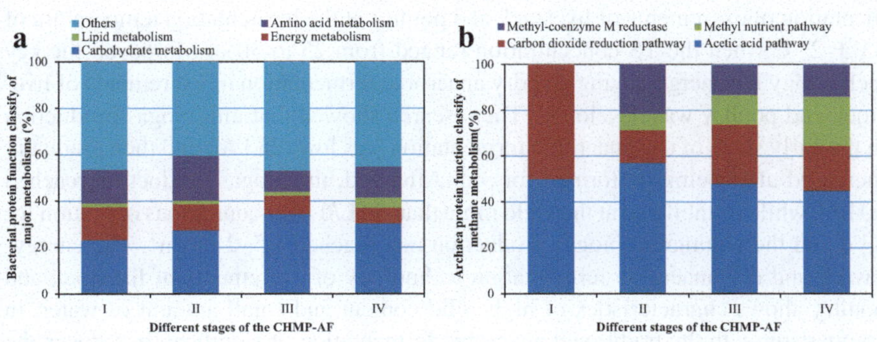

Fig. 2.26 Protein function classify of bacterial major metabolisms and archaea methane metabolism in different stages of the CHMP-AF (Jia et al. 2017d)

archaeal proteins. Eventually, methane was produced due to the effects of methyl-coenzyme M reductase. The results revealed that numerous proteins of archaea and bacteria participated in the production of methane by using the reduction pathway of CO_2, mainly including bacteria of *Azotobacter* and archaea of *Methanoculleus, Methanosarcina, Methanocaldococcus* and *Methanothermobacter* (Jia et al. 2017d).

2.3.4 Dry Anaerobic Fermentation Technology of Livestock Manure

In recent years, livestock and poultry breeding industries in villages and towns in China have changed from the traditional courtyard-type free breeding with small quantities to field centralized-type with large quantities. However, peasant households are still the unit of the main breeding subjects and non-point source pollution is serious. These livestock and poultry breeding industries are small and highly decentralized, have mixed compositions of excrements of livestock and poultry and significant regional differences. Aiming at this, key manageable techniques and equipment of low pollution, low cost, and low maintenance requirement for dry anaerobic fermentation of excrements of livestock and poultry were developed. The purpose is to improve energy-fertilizer integrated utilization of wastes in decentralized breeding.

2.3.4.1 Biogas Production Performance

Dry anaerobic fermentation is an anaerobic fermentation technology in which the concentration of TS in the fermentation liquid exceeds 20%. Due to too high TS concentration, continuously or semi-continuously feeding is difficult and the bulk feeding is adopted in most situations (Fig. 2.27). Cow dung was used as research object in which anaerobic digested sludge was inoculated (Figs. 2.28, 2.29 and 2.30). Based on this, this study investigated biogas production performances of dry anaerobic fermentation of excrements of livestock and poultry at the fermentation temperature of 35 ± 2 °C when the TS concentration ranged from 25 to 30%. Moreover, the key technology for energy-saving and dry anaerobic fermentation of excrements of livestock and poultry was developed. The research showed that unit biogas production in the early stage of dry anaerobic fermentation was low (25 L/d) and then gradually increased after being performed for 3 d. After 8 d, the biogas production reached 50 L/d, while it stabilized at the yield more than 50 L/d after continuous operation for 43 d and the volumetric biogas production rate reached 1.5–2 m^3/m^3. The energy-saving and dry anaerobic fermentation technology of excrements of livestock and poultry shows characteristics of high solid content and small amount of water. In comparison with the traditional anaerobic fermentation, it significantly reduces the volume of materials participating in reaction and investment, saves more than 30% of energy consumption for operation and has no secondary pollution.

Fig. 2.27 Simulation device of dry anaerobic fermentation

2.3.4.2 Control for Process Operation

The compositions of organic wastes produced in livestock and poultry breeding are complex. In view of this, this study established the strategies for optimising and controlling dry anaerobic fermentation process of excrements of livestock and poultry in villages and towns based on the researches on the technologies, such as material ratio, parameters of fermentation process and process control. Moreover,

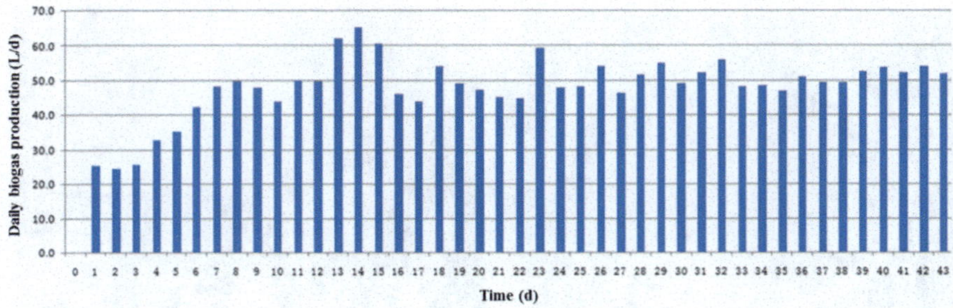

Fig. 2.28 Changes of biogas production in dry anaerobic fermentation

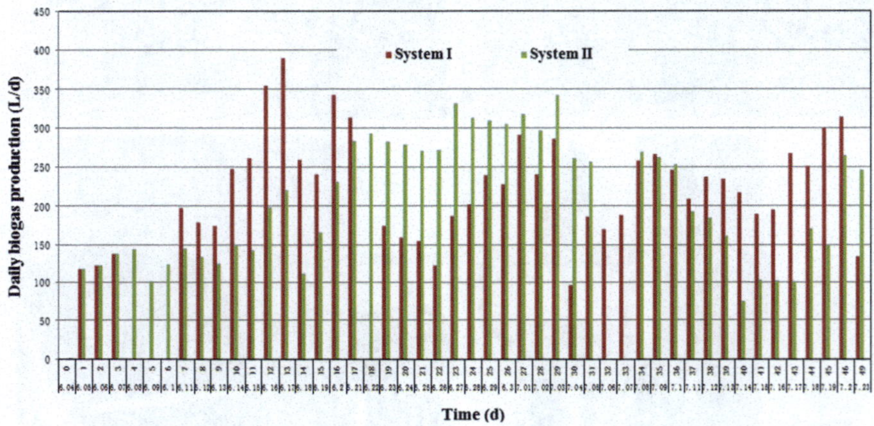

Fig. 2.29 Changes of daily biogas production of different processes

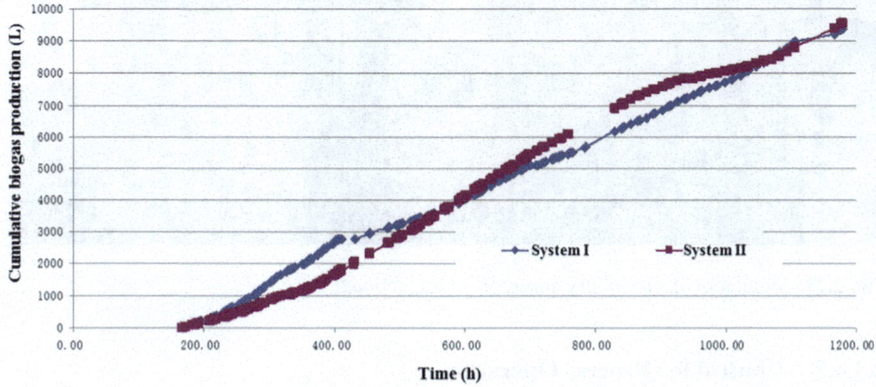

Fig. 2.30 Changes of cumulative biogas production of different processes

this research developed small and automatic dry anaerobic fermentation equipment suiting for villages and towns.

Based on the research of biogas production performance of dry anaerobic fermentation of excrements of livestock and poultry, the influences of feed amount and reflux ratio on biogas production of dry anaerobic fermentation of excrements of livestock and poultry were studied. Furthermore, this study performed the research on quick start and control over stable operation of the dry anaerobic fermentation system (Figs. 2.31 and 2.32). By taking fresh cow dung as raw materials, the materials were fed after being uniformly mixed with different feed amounts (5, 8 and 10 L) at different reflux ratios (1:1 and 2:1). Based on this, changes of biogas production and cumulative biogas production per unit time were compared and analysed. At the beginning of starting equipment, feeding load of system 1 increased from 5 to 8 L/d, while that of system 2 kept at 5 L/d. In the process of rising the reflux ratio from 1:1 to 2:1, the biogas production per unit time of the system 1 was significantly higher than that of system 2. This indicated that improving feed loading had certain promotion to biogas production rate in the early stage of start. After further increasing feed loading of the system 1 to 10 L/d, the biogas production per unit was lower than that of the system 2. At this time, the feed mode of the system 2 was feed twice a day with 5 L of feed loading for each at reflux ratio of 2:1. The results of cumulative biogas production demonstrated that feed amount of the system 1 was obviously larger than that of the system 2, while the increase of cumulative biogas production was not obvious. In the middle and late stage of operation, the cumulative biogas production of the system 2 was higher than that of the system 1. Obviously, biogas production effect of dry anaerobic fermentation is closely related to feed mode and the best biogas production effects can be obtained when the feed loading is 10 L/d and reflux ratio is 1:1.

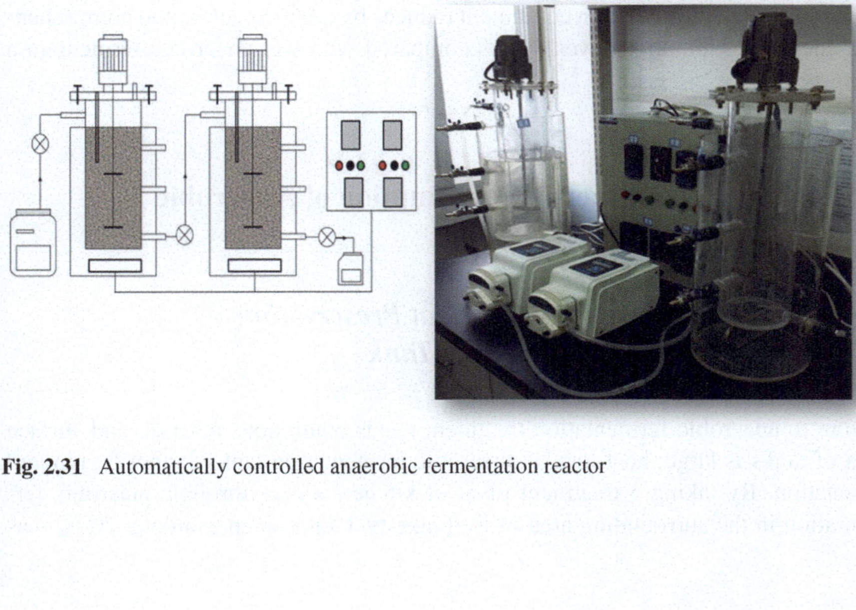

Fig. 2.31 Automatically controlled anaerobic fermentation reactor

Fig. 2.32 Small, vertical and energy-saving dry fermentation equipment

Based on the operation control of dry anaerobic fermentation process, energy consumption of the process was evaluated from the aspects of biogas production, unit energy consumption and total energy consumption. In comparison with the traditional wet anaerobic fermentation equipment with the same biogas production scale, the unit biogas production of dry anaerobic fermentation was only 22.25% that of wet anaerobic fermentation. Moreover, unit energy consumption of products of dry anaerobic fermentation equipment reduces by 0.069 kgce/m^3 and comprehensive energy consumption saves 34.8% compared with wet anaerobic fermentation equipment.

2.4 Evaluation of Energy Consumption of Anaerobic Fermentation Process

2.4.1 Energy Consumption of Heat Preservation of Anaerobic Fermentation Tank

Owing to anaerobic fermentation treatment needs continuous reaction and surface area of tanks is large, heat loss of anaerobic fermentation tanks cannot be ignored in reaction. By taking a treatment plant of kitchen wastes through anaerobic fermentation in the surrounding area of Beijing city, China as an example, 20 °C was used as the local average temperature and anaerobic fermentation process at the constant moderate temperature of 35 °C was adopted. A continuous stirred tank reactor (CSTR) was used as the fermentation tank and the volume of a single-phase reaction tank was 500 m^3. In the CHMP-AF process, the volume of the tank for hydrogen production was 50 m^3, while the volume of the tank for methane production was 450 m^3. Moreover, height-diameter ratio of all tanks was 5:4. The tanks were manufactured by using splicing technology of steel plates and rock wool was used in interlayer of the tanks for heat preservation. Heating coils were added on internal walls for keeping the constant temperature of anaerobic fermentation liquids. Owing to thermal resistance ($R = 3.45 \times 10^{-4}$ m^2 °C/W) of steel plates was very small and far smaller than that of rock wool ($R = 4.2$ m^2 °C/W), it can be neglected. The thermal energy for heat preservation of tanks is calculated according to the following formula.

$$E_J = S_J \times \Delta Tem \times \Delta Time/R \tag{2.1}$$

$$E_L = S_L \times \Delta Tem \times \Delta Time/R \tag{2.2}$$

$$S_L = S_H + S_C \tag{2.3}$$

where, E_J represents the total energy consumption of heat preservation of tanks in the single-phase methane-production process, E_L indicates the total energy consumption of heat preservation of tanks in the combined production process. S_J and S_L stand for the surface areas of tanks for single-phase methane production and combined production separately. S_H and S_C denote the surface areas of tanks for hydrogen and methane production, respectively. $\Delta Tem = 15$ °C shows the differences between ambient temperature and constant temperature of anaerobic fermentation. $\Delta Time = 1\ d$ demonstrates the unit time of anaerobic fermentation process and it is 1 d. R represents the thermal resistance of rock wool and $R = 4.2$ m^2 °C/W.

Through the calculation, the surface area of tank for single-phase process of methane production is $S_J = 351.68$ m^2, while that of the two-phase reactor for combined production process is $S_L = 402.53$ m^2. In addition, $E_J = 30.144$ KW h and $E_L = 34.503$ KW h.

2.4.2 *Energy Consumption in Feeding and Heating*

Assuming that one ton of kitchen wastes are treated by the treatment plant of anaerobic fermentation every day and ambient temperature is 20 °C, hydrothermal pre-treatment is performed at 90 °C and residual heat of kitchen wastes after hydrothermal pre-treatment is recovered. A plate heat exchanger with the heat transfer efficiency of $\eta = 80\%$ is used (Li et al. 2014). Therefore, the calculation formula for energy consumption in feeding and heating is shown as follows.

$$Q = C \times M \times \Delta T \tag{2.4}$$

where, C represents the specific heat of materials. Water content of kitchen wastes maintains at about 95% and to keep stable operation of the reaction system, influent load is low, and influent water can be regarded as homogeneous water phase. Therefore, C is calculated according to specific heat of water and C = 1 cal/g°C. M indicates the mass of materials fed and measured in unit of g. ΔT is the difference between ambient temperature and constant temperature of anaerobic fermentation. When performing hydrothermal pre-treatment, ΔT is the difference between ambient temperature and set temperature for hydrothermal pre-treatment.

Based on the calculation, the energy consumption in feeding and heating of kitchen wastes without undergoing hydrothermal pre-treatment is $Q_Y = 1.5 \times 10^7$ cal, while that of hydrothermally pre-treated kitchen wastes is $Q_S = 7 \times 10^7$ cal. By using a heat exchanger, residual heat in hydrothermal pre-treatment of kitchen wastes can be recovered and $Q_R = 4.4 \times 10^7$ cal.

2.4.3 *Energy Output in Anaerobic Fermentation*

The density of hydrogen is $\rho_{H2} = 0.0899$ g/L and its heat value is $q_{H2} = 143$ kJ/g. The density and heat value of methane are $\rho_{CH4} = 0.717$ g/L and $q_{CH4} = 55.9$ kJ/g, respectively.

$$Q_{H2} = \rho_{H2} \times q_{H2} \times V_{H2} \tag{2.5}$$

$$Q_{CH4} = \rho_{CH4} \times q_{CH4} \times V_{CH4} \tag{2.6}$$

$$Q_{JS|JY|LS|LY} = Q_{CH4} + Q_{H2} \tag{2.7}$$

Heat of different units is conversed as follows:

1 cal = 4.1868 J.

1 KWh = 3.6 × 106 J.

Energy consumption in heat preservation of anaerobic fermentation tanks per unit time is $E_J = 1.08 \times 10^8$ J and $E_L = 1.24 \times 10^8$ J and energy consumption in feeding and heating is $Q_S = 2.933 \times 10^8$ J, $Q_Y = 0.629 \times 10^8$ JandQ$_R = 1.84 \times 10^8$ J. The total energy and biogas production capacity of anaerobic fermentation are

$$E_{JS} = Q_{JS} - E_J - Q_S + Q_R \tag{2.8}$$

$$E_{JY} = Q_{JY} - E_J - Q_Y \tag{2.9}$$

$$E_{LS} = Q_{LS} - E_L - Q_S + Q_R \tag{2.10}$$

Table 2.6 Energy flow analysis of different anaerobic fermentation processes

	SHP-S	NSHP-S	SHP-T	NSHP-T
Hydrogen production ratio (VS L/g)	–	–	0.14	0.11
Methane production ratio (VS L/g)	1.73	1.66	3.52	2.64
Hydrogen production (L/d)	–	–	1400	1100
Methane production (L/d)	17,300	16,600	35,200	26,400
Hydrogen (g)	–	–	125.86	98.89
Methane (g)	12,404.1	11,902.2	25,238.4	18,928.8
Hydrogen heat gain (KJ/d)	–	–	17,997.98	14,141.27
Methane heat gain (KJ/d)	693,389.2	665,333	1,410,827	1,058,120
Total heat (KJ/d)	693,389.2	665,333	1,428,825	1,072,261

$$E_{LY} = Q_{LY} - E_L - Q_Y \tag{2.11}$$

where, E_{JS}, E_{JY}, E_{LS} and E_{LY} represent the net energy income through JS, JY, LS and LY treatments, respectively.

According to one ton of daily feed amount, biogas production obtained under four different processes is shown in Table 2.6. Through conversion, it can be obtained that $E_{JS} = 4.757 \times 10^8 J$, $E_{JY} = 4.94 \times 10^8 J$, $E_{LS} = 11.967 \times 10^8 J and E_{LY} = 8.83 \times 10^8 J$. Obviously, the highest net energy income can be obtained through the CHMP-AF process for kitchen wastes undergoing hydrothermal pre-treatment at 90 °C.

2.5 Research and Development of Equipment

2.5.1 Simulation Device for Combined Hydrogen-Methane Production

In view of the needs of simulation test of anaerobic fermentation, an automatically controlled simulation device with the volume of 10 L for combined hydrogen-methane production was developed. This device controlled the feeding rate through a peristaltic pump and regulated operation parameters, such as pH value, temperature and stirring speed through numerical control devices. By using this device, the research on CHMP-AF of organic wastes was conducted, which provides technical supports for pilot plant tests and demonstration projects.

2.5.2 Small and Vertical Dry Fermentation Equipment

In view of decentralised organic wastes in villages and towns, small and vertical dry fermentation equipment with the capacity of 500 L was developed. This equipment comprised a tank, a feed inlet, an outlet, a temperature controller, thermal insulation layers, an electric heat tracing and biogas pipes. The fermentation temperature is 35 ± 1 °C and feed amount is controlled in the range of 25–30%, which could realise high-efficient dry anaerobic fermentation of organic wastes. This equipment shows the characteristics, such as occupying a small area, low operation cost and simple operation, so it suits for energy regeneration from organic wastes, like excrements of livestock and poultry, kitchen wastes and surplus sludge in small and decentralised breeding plants in villages and towns.

2.5.3 Horizontal Energy-Saving Dry Fermentation Equipment

As for agricultural organic wastes, such as excrements of livestock and poultry and straws, a small, horizontal and energy-saving fermentation equipment with the total volume of 6 m^3 and effective volume of 4 m^3 was developed (Fig. 2.33). The feed amount is controlled between 25 and 30% and the fermentation temperature is 35 ± 2 °C. Due to the advantages, such as occupying small areas, low energy consumption in operation and less secondary pollution, it is suitable for energy utilization of organic wastes in villages and towns.

Fig. 2.33 Horizontal energy-saving fermentation equipment

2.5.4 Container-Type Anaerobic Fermentation Equipment

Aiming at organic wastes, such as kitchen wastes and wastes produced in pedlars' markets, a set of standard, modular and integrated on-board container-type anaerobic fermentation equipment was developed (Fig. 2.34). The modules, like a feeding system, a discharge system, a fermentation tank and an on-line monitoring system of biological gases were included in the equipment and integrated in the container. This realises modularisation of process parameters, automation of equipment operation and data collection and remote control under the platform of Internet of things. This

Fig. 2.34 On-board container-type integrated anaerobic fermentation equipment

equipment suits for energy regeneration of decentralised and small-scale organic wastes in villages and towns for the characteristics, such as occupying small areas, mobility, stable operation and simple operation.

References

Jia X, Li M, Xi B, et al. (2014). Integration of fermentative biohydrogen with methanogenesis from fruit–vegetable waste using different pre-treatments. Energy Conversion and Management, 88:1219–1227.

Jia X, Zhu C, Li M, et al. (2013). A comparison of treatment techniques to enhance fermentative hydrogen production from piggery anaerobic digested residues. International Journal of Hydrogen Energy, 38(21):8691–8698.

Jia X, Xi B, Li M, et al. (2017a). Evaluation of biogasification and energy consumption from food waste using short-term hydrothermal pretreatment coupled with different anaerobic digestion processes. Journal of Cleaner Production, 152:364–368.

Jia X, Xi B, Li M, et al. (2017b). Metaproteomic analysis of the relationship between microbial community phylogeny, function and metabolic activity during biohydrogen-methane coproduction under short-term hydrothermal pretreatment from food waste. Bioresource Technology, 245(Pt A):1030–1039.

Jia X, Xi B, Li M, et al. (2017c). Biodegradation of Potamogeton pectinatus biomass by hydrogen and methane coproduction: Effect of different pretreatments and parallel factor analysis. International Journal of Hydrogen Energy, 42(29):18315–18324

Jia X, Xi B, Li M, et al. (2017d). Metaproteomics analysis of the functional insights into microbial communities of combined hydrogen and methane production by anaerobic fermentation from reed straw. PLoS ONE, 12(8):e0183158.

Li M, Xia T, Zhu C, et al. (2014). Effect of short-time hydrothermal pretreatment of kitchen waste on biohydrogen production: Fluorescence spectroscopy coupled with parallel factor analysis. Bioresource Technology, 172:382–390.

Chapter 3
Bio-augmented Composting of Organic Wastes in Villages and Towns

Abstract Composting technology of wastes has been applied in Chinese farming activities long before while it was not scientifically discussed until 1920. Composting is a method of decomposing, maturing and transforming complex soil organic matters into stable similar humus soil through manual control under the biochemical action of microorganisms. The method can be divided into thermophilic aerobic and low-temperature anaerobic composting according to the mechanism of bacteriolysis. Based on composting mode, there are open-air composting and mechanical composting. The composting process plays an important part in safely treating the organic wastes and improving soil fertility. Composting can also provide nutrients for plant growth and increase the capability of soil for retaining minerals.

3.1 Overview

3.1.1 Composting Process

The composting process is divided into four steps: (1) pre-treatment. In the step, massive inorganic impurities are eliminated and then the wastes are crushed and screened to attain homogeneous particles which show optimal moisture content of 45–60% and the C/N ratio about (20–30): (1) On the condition of not satisfying the requirement, sludges or feces can be doped; (2) fermentation. On the condition of having suitable temperature, moisture and oxygen, aerobiotic or anaerobic microorganisms are rapidly multiplied, causing organic matters to be decomposed to thus further result in the transformation of various matters into harmless fertilizer; (3) maturation. The step keeps the quality of fertilizer stable, and the fertilizer can be applied after being fully matured; (4) storage or disposal. The acquired fertilizers are stored while waste fertilizers are subjected to landfill disposal.

Figure 3.1 displays the air motion in static compost pile. Heat is generated due to the degradation of organic matters in the pile. Therefore, hot air flows upwards the upper part of the pile while the fresh air flows into the pile from the bottom to thus form convection, guaranteeing the oxygen supply of the pile. Additionally, the wind and diffusion effect also can bring a part of oxygen for the pile.

M. Li et al., *Differentiated Resourceful Utilization of Rural Organic Wastes*,
https://doi.org/10.1007/978-981-15-2712-8_3

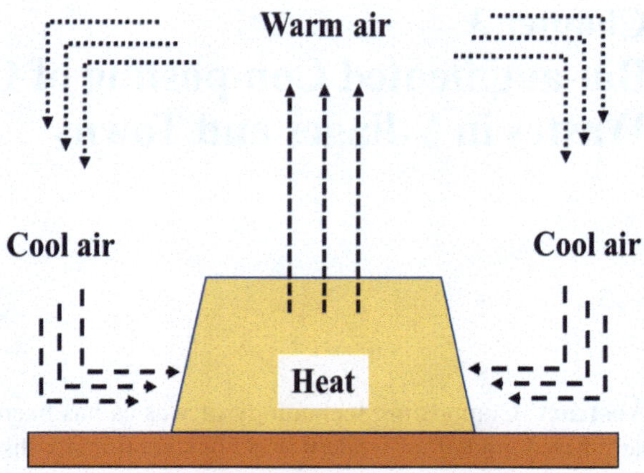

Fig. 3.1 Convection currents and other factors affecting aeration of a static compost pile

According to temperature change, the composting process can be divided into different stages. Figure 3.2 reveals the typical change curves of temperatures during static-pile composting and forced-aeration composting. The optimum growth temperature of bacterial community is different from that of fungal community. For example, the optimum growth temperatures of psychrophilic bacteria, mesophilic bacteria and thermophilic bacteria are 13 °C, 21–48 °C and 45–68 °C, respectively. On this basis, the composing process can be partitioned into temperature-rising, high-temperature and maturation periods. During the temperature-rising period of composting, the temperature of the pile gradually increases, then rises to the peak at the high-temperature period and finally gradually declines to enter the maturation period.

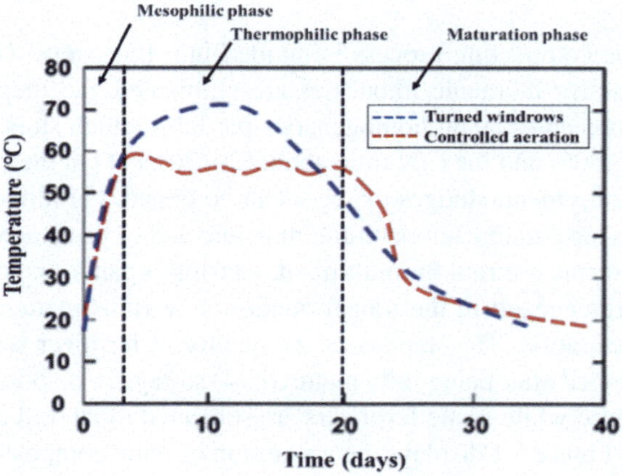

Fig. 3.2 Typical temperature curves for static composting and aerated composting

3.1.2 *Influencing Factors of Composting Process*

3.1.2.1 Organic Matters

The content of organic matters influences the temperature of windrows and ventilation and oxygen supply. At extremely low content of organic matters, the heat generated due to decomposition is insufficient to promote and sustain proliferation of thermophilic bacteria during fermentation. Therefore, the temperature of compost piles is hard to reach the high-temperature period, which influences the sanitary and non-hazardous effect of fermentation. Moreover, extremely low content of organic matters can influence the fertilizer efficiency and utilization value of fermentation products. However, under an extremely high content of organic matters, it is necessary to provide a great oxygen supply, which practically brings difficulties for oxygen supply during the turning. It is likely to result in partially anaerobic condition due to lacking of oxygen supply. The moderate content of organic matters is in the range of 20–80%.

3.1.2.2 C/N Ratio

C/N ratio is one of important parameters for investigating the composition of compost mixture. If the C/N ratio is too high, the temperature of compost piles likely fails to rise; otherwise, the compost probably emits malodorous gas. The initial C/N ratio of the compost mixture is supposed to be ranged from 25:1 to 40:1, and even 50:1. However, different materials made of wood fibres show a great difference in density, particle size and lignin content and therefore the specific range of initial C/N ratio is hard to define. With the increase of C/N ratio, the fermenting time relatively prolongs. When the C/N ratio of the raw materials is 20, 30–50, and 78, the corresponding required fermenting time is about 9–12, 10–19 and 21 days, respectively. However, when the C/N ratio is larger than 80:1, fermentation is difficult to appear. The C/N ratio decreases during the composting process and generally maintains at (10–20):1 in the maturation stage. The change of C/N ratio during maturation and fermentation of materials can be considered as a criterion for maturation of compost products, bringing favorable fertilizer efficiency during agricultural utilization.

3.1.2.3 Moisture Content

Moisture is crucial for the composting process, which can promote degradation of substrates by moving active microorganisms and offer a favorable condition for nitrogen fixation. Whether moisture is proper directly influences the fermenting speed and maturation. A high humidity can block small pores and further restricts the transport of oxygen, thus forming anaerobic condition to release odorous gas. Therefore, the moderate moisture content for the compost pile is in the range of 55–65%. In practical application, the simple method to make judgment is to tightly hold the materials in hands to find water between fingers while water does not drop.

3.1.2.4 Particle Size

The oxygen for fermentation is supplied through pores in particles of fermenting materials. The porosity and pore size depend on the particle size and structural strength. Paper, animals, plants, fabrics, etc. can increase their densities when they are in water under compression and pores between their particles greatly decline, which is unfavorable for ventilation and oxygen supply. The suitable size of particles is generally in the range of 12–60 mm.

3.1.2.5 Turning

Turning for oxygen supply is one of fundamental conditions of production of aerobic composting. The purpose of turning is to provide oxygen in order to further accelerate the fermentation process of microorganisms, adjust the temperature of the pile and dry the windrows. If the times of turning are low, the ventilation rate is low so that there is no sufficient oxygen for microorganisms, restricting the increase of fermentation temperature. However, excessive turning times can cause the heat loss of the windrows, which influences the non-hazardous degree of fermentation. Generally, it is necessary to turn the compost piles for 2–3 times during the fermentation based on actual condition.

3.2 Development of Complex Microbial Inoculants

3.2.1 Complex Microbial Inoculants for Degrading Lignocelluloses

The microbial community in mature compost was taken as materials and the bran rich in cellulose was used as carbon source to thus promote the proliferation of cellulose-decomposing bacteria and induce microorganisms to generate ligninase. In this way, efficient complex microbial inoculants for degrading lignocelluloses can be prepared.

Culture of the complex microbial inoculants: Ten grams of mature compost was added into 1000 mL of peptone cellulose solution (PCS) medium to statically culture the complex microbial inoculants at 50 °C for 7 days. The PCS medium was prepared with 5 g of peptone, 5 g of NaCl, 2 g of $CaCO_3$, 1 g of yeast powder, 5 g of bran and 1000 mL of deionized water and sterilized for 20 min at 121 °C. Moreover, 10 mL of bacterium suspension was added into 1000 mL of PCS medium for conducting subculture, which was continuously carried out for ten generations in order to stabilize the microbial community structure.

Isolation and identification of complex microbial inoculants: By using spread plate method, 20 strains of bacteria and 10 trains of fungi were separately screened from the inoculation agent. By using boiling method, minor DNA was rapidly extracted from bacteria for polymerase chain reaction (PCR) amplification while genomic DNA of fungi was extracted by employing DNA extraction kit. Furthermore, 5 μL of extracting solution for genome of fungi was taken and detected by applying 1% of agarose gel electrophoresis (AGE), showing a clear and bright band, indicating that no degradation appeared. Therefore, PCR amplification of 28S rDNA gene can be conducted. After conducting PCR amplification of 16S rDNA gene, target fragments of 16S rDNA genes of 20 strains of bacteria were acquired, with the size about 1600 bp obtained through AGE, as shown in Fig. 4.5. After carrying out PCR amplification of 28S rDNA genes, only the target fragments of 28S rDNA gene of a strain of fungus were attained, with the size about 700 bp detected through AGE.

The purified PCR products were sequenced and the result is displayed in Table 3.1. To be specific, 20 strains of bacteria were divided into eight types, including *Aeribacillus* sp., *Bacillus licheniformis*, *Bacillus thermoamylovorans*, *Brevibacillus borstelensis*, *Geobacillus* sp., *Lysinibacillus* sp., *Paenibacillus dendritiformis* and *Ureibacillus thermosphaericus*, all belonging to Bacillales. The fungus *Penicillium* sp. belongs to Eurotiomycetes.

3.2.2 Acidification-Resistant Complex Microbial Inoculants

Culture of complex microbial inoculants: During inhibitory stage of acidification of kitchen waste compost, some samples were taken to prepare bacterium suspension. By using limiting culture and oriented acclimatization generation of culture medium of mixed acids [m (acetic acid: propionic acid: butyric acid: lactic acid)], acidification-resistant complex microbial inoculants capable of efficiently degrading micromolecular organic acids with were prepared. The culture was conducted using complete medium (50 mL), and the proportion of mixed acids was 3:3:2.5:12.5 and pH was adjusted to 5.3, in which 1 mL of bacterium suspension was added. The mixture was cultured at the constant temperature of 50 °C for 24 h at the rotational speed of 200 r/min. According to the aforementioned steps, subculture was carried out in which addition amount of mixed acids increased step by step. After continuously culturing 50 generations, microbial community structure was stable, that is, acquiring a complex microbial system for degrading organic acids.

Isolation and identification of complex microbial inoculants: The SEM images, colonial morphology and characteristics of seven strains of single bacteria isolated from acidification-resistant complex microbial inoculants are shown in Fig. 3.3. It can be seen from the Fig. 3.3 and Table 3.2 that round light yellow S-1 bacterial colony with the diameter about 5 mm was transparent, whose surface was moist,

Table 3.1 Phylogenetic identification of the bacterial isolates from the inoculation agent

Bacterial isolate	Accession number	Closest match	Similarity (%)	Class/order
1	KC441060	*Aeribacillus* sp.	99	*Bacilli, Bacillales*
2	KF952566	*Brevibacillus borstelensis*	100	*Bacilli, Bacillales*
3	GQ375243	*Bacillus licheniformis*	99	*Bacilli, Bacillales*
4	KF952564	*Ureibacillus thermosphaericus*	99	*Bacilli, Bacillales*
5	AB211001	*Ureibacillus thermosphaericus*	100	*Bacilli, Bacillales*
6	AB211001	*Ureibacillus thermosphaericus*	99	*Bacilli, Bacillales*
7	HM071942	*Paenibacillus dendritiformis*	100	*Bacilli, Bacillales*
8	AB360808	*Bacillus thermoamylovorans*	99	*Bacilli, Bacillales*
9	JX312637	*Lysinibacillus* sp.	99	*Bacilli, Bacillales*
10	AB682456	*Ureibacillus thermophilus*	99	*Bacilli, Bacillales*
11	AB682456	*Ureibacillus thermophilus*	99	*Bacilli, Bacillales*
12	NR_043747	*Ureibacillus thermophilus*	99	*Bacilli, Bacillales*
13	HM059720	*Geobacillus* sp.	99	*Bacilli, Bacillales*
14	NR_113799	*Brevibacillus borstelensis*	98	*Bacilli, Bacillales*
15	AB211001	*Ureibacillus thermosphaericus*	99	*Bacilli, Bacillales*
16	AB211001	*Ureibacillus thermosphaericus*	99	*Bacilli, Bacillales*
17	AB360816	*Bacillus thermoamylovorans*	99	*Bacilli, Bacillales*
18	AB360815	*Bacillus thermoamylovorans*	99	*Bacilli, Bacillales*
19	KC441060	*Aeribacillus* sp.	100	*Bacilli, Bacillales*
20	FN666260	*Bacillus thermoamylovorans*	99	*Bacilli, Bacillales*

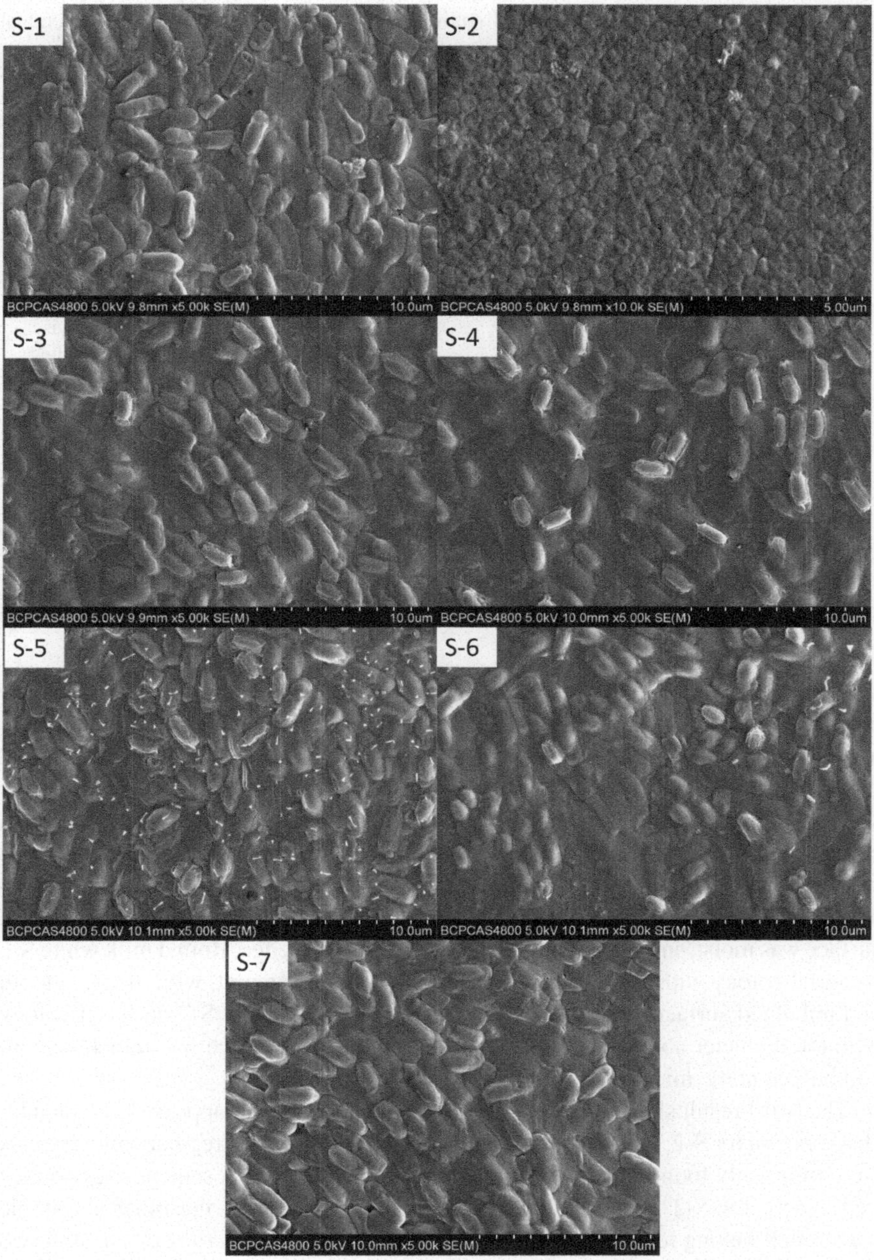

Fig. 3.3 The SEM photographs of 7 isolates of microbial consortium degrading organic acids

Table 3.2 The characteristics of colonial morphology of pure culture isolates from microbial consortium degrading organic acids

Serial number of bacterial strains	Shape of bacterial colony	Color	Transparent	State of moisture content	Having folds or not	Having bulges or not	Edge morphology
S-1	Round	Light yellow	Y	Moist	N	N	Ringwise extending
S-2	Round	Yellow	N	Moist	N	N	Regular
S-3	Round	Light yellow	N	Moist	N	Y	Ringwise extending
S-4	Round	Milk white	N	Moist	N	N	Incised shape
S-5	Round	Milk white	N	Moist	N	N	Regular
S-6	Round	Milk white	N	Moist	N	N	Regular
S-7	Round	Light yellow	Y	Moist	N	Y	Incised shape

smooth and not bulged and its edge was ringwise extended. Yellow S-2 bacterial colony was small and non-transparent and appeared as spots, with irregular shape and moist, smooth and bulged surface. Moreover, its edge was regular. Round light yellow S-3 bacterial colony with the diameter about 6 mm was non-transparent. Its surface was moist and did not show folds while was bulged, and its edge was extended ringwise. Round milk white S-4 bacterial colony was non-transparent, and its surface was moist, smooth and not bulged. Its edge appeared as an incised shape. Round milk white S-5 bacterial colony with the diameter of 2–3 mm was non-transparent and its surface was moist, smooth and not bulged, with a regular edge. Round milk white S-6 bacterial colony with 4–5 mm in diameter was non-transparent, with moist, smooth and unbulged surface and a regular edge. Round light yellow S-7 bacterial colony with the diameter about 4 mm was transparent, and the surface was moist, smooth and bulged at the middle part, showing an incised edge.

The SEM result showed that the single bacterial strains all appeared as a slender shape except for S-2, which appeared as an irregular ellipsoid. Irregular white crystals were frequently found on S-5 bacteria. The result of 16Sr DNA sequencing revealed that except for S-2, which belongs to *Microbacterium* spp., the other six single bacteria all belong to *Bacillus* spp. in which there are one *Bacillus cereus* and two *Bacillus licheniformis*. As shown in Table 3.3, the seven single bacterial strains are all subordinated to acetic acid- and propionic acid-degrading microorganisms.

Table 3.3 Sequencing result of single bacteria by using plate isolation method

Serial number	Accession number	The species with the greatest similarity	Similarity (%)
S-1	GU250442.1	*Bacillus cereus* strain BFE 5400	99
S-2	AB847928.1	*Microbacterium* sp. ZYZR2	99
S-3	HE804782.1	*Bacillus licheniformis*	99
S-4	EU593762.1	*Bacillus* sp. 00763	99
S-5	JF937058.1	*Bacillus* sp. JSG1	98
S-6	JF701942.1	*Bacillus* sp. YXA3-25	99
S-7	EU257697.1	*Bacillus licheniformis strain DC3-1*	99

3.3 Biological Bio-augmented Composting and Secondary Pollution Control Technology

3.3.1 Bio-augmented Composting of Municipal Solid Wastes

3.3.1.1 Experimental Design and Operation

Research showed that during composting, inoculating efficient microbial inoculants can increase the number of microorganisms in piles, speed up the decomposition of organics in wastes, promote the maturation of composting materials and further improve the composting efficiency. Some scholars suggested that the number and category of microbial community in composting system were enough. Therefore, the microbial community can be rapidly propagated under proper conditions to further quickly start composting by stimulating indigenous microorganism. The research explored the effect of lignocelluloses-degrading inoculants in promoting the ligno-cellulose degradation and the humification process of compost by inoculating the lignocelluloses-degrading inoculants.

During the research, two experimental groups (with and without addition of microbial inoculants) were designed. For the experimental group added with microbial inoculants, the addition amount (dry weight) of microbial inoculants was 5 mL/kg. The ratio of addition amounts (dry weight) of kitchen wastes, humus soil and bran was 15:7:1. Additionally, sawdust was added as filler in order to improve the pore structure of the pile and increase the ventilation effect. The initial moisture content of materials in the compost was adjusted to about 60% while the ventilation rate was 0.5 L/min kg. After separately being stacked up for 5, 15 and 30 days, the samples were taken for analysis of theoretical index and molecular biology.

3.3.1.2 Promotion Mechanism of Lignocelluloses-Degrading Microbial Inoculants for Humification of Composts

By using metaproteomics method, the influence of inoculating mixed bacteria of lignocelluloses on structures and functions of bacterial and fungal communities in the composting process of kitchen wastes was analyzed (Fig. 3.4). Moreover, the differences of metabolic pathways of carbohydrates during the composting were further analyzed to thus further reveal the action mechanism of inoculants on microorganisms in the composting process. Metaproteomics research showed that compared with the group without being inoculated, after inoculating the cellulose-degrading microbial inoculants, the proportion of bacteria in *Pseudomonadales* rose from 15.1 to 27.6% while that of Bacilli decreased from 16.0 to 6.4%. Additionally, the proportion of *Eurotiomycete* fungi increased to 32.8 from 10.8% while that in *Saccharomycete* fungi reduced from 48.1 to 26.6%. Adding microbial inoculants showed an insignificant influence on the diversity of microbial community in composts while the number of microbial communities varied. Moreover, *Pseudomonadales* became the predominant bacteria by replacing Bacilli while *Eurotiomycetes* was taken as predominant fungus by replacing *Saccharomycetes*.

The metabolism of carbohydrates is the major metabolic type in the composting process. After inoculating cellulose-degrading microbial inoculants, the bacteria in *Pseudomonadales* (17) and *Enterobacteriales* (12), *Actinomycetes* (6) in *Actinobacteria* and fungi (9) in *Eurotiomycetes* were more than those of the control group. By contrast, the numbers of Bacilli bacteria (10) and *Saccharomycetes* fungi (12) in the control group were larger.

In carbohydrates, the degradation of lignocelluloses plays a crucial role in the composting process. As the main carbon sources, cellulose and hemicelluloses provide energies for biotransformation and lignin is the important starting material during the formation of humic acid. Nine lignocellulolytic enzymes were identified from the experimental group (addition with microbial inoculants) and the control group. In the experimental group with the addition of microbial inoculants, there were cellulase (1) generated by *Thermobifida* spp., cellulase (4) and hemicellulase (1) produced by fungi in *Aspergillus* (*Eurotiomycetes*), and ligninase (1) produced by fungi in *Melanocarpus* (*Sordariomycetes*). The cellulase (1) generated by bacteria in *Bacillus* (Bacilli) and hemicellulase (1) produced by fungi in *Aspergillus* were detected in the control group, as shown in Table 4.2. It indicated that during the experimental process with addition of microbial inoculants, fungi in *Aspergillus* (*Eurotiomycetes*) and *Melanocarpus* are the main microorganisms for degrading lignocelluloses in the composting process while the Basidiomycota fungi are both weak in abundance and activity.

As shown in Fig. 3.5, bacteria in *Bacillales* and *Aspergillus* spp. in microbial inoculants were the main microbial communities in which *Aspergillus* spp. can degrade

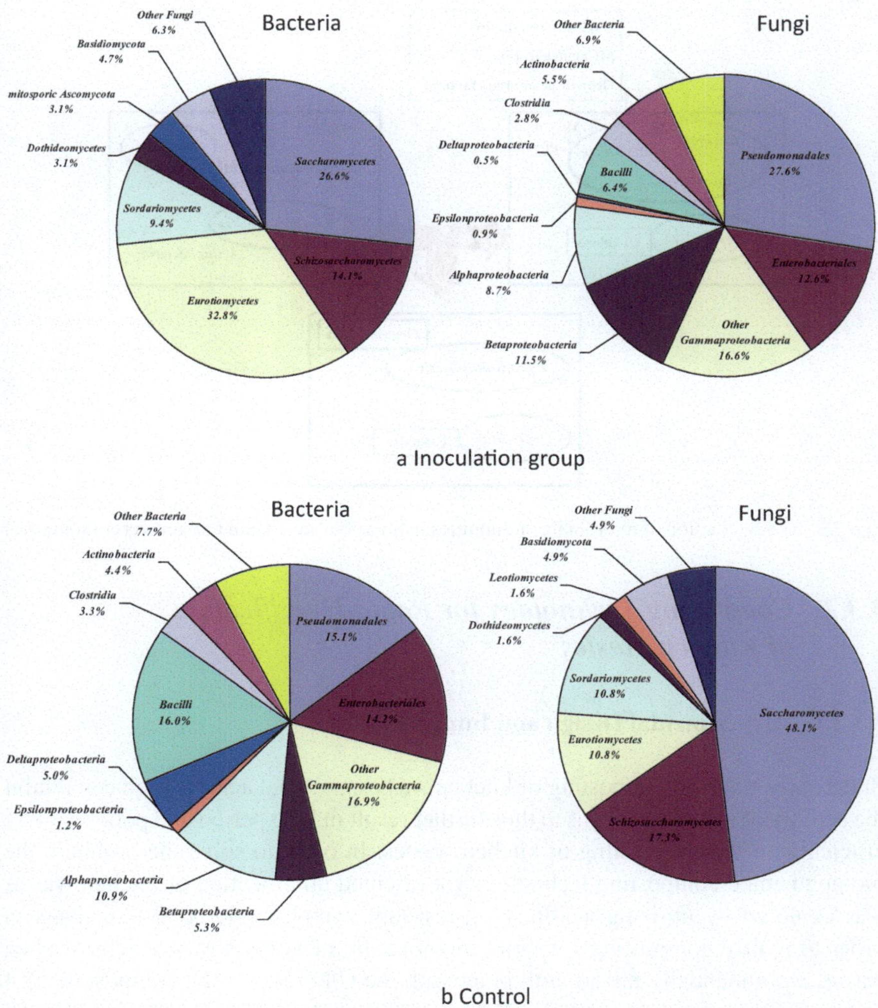

Fig. 3.4 Phylogenetic assignment of bacterial and fungal proteins detected in the metaproteome

cellulose while bacteria of *Bacillales* can stabilise the system. The bacteria of *Bacillales* and fungi of *Saccharomycetes* are the major communities in the compost. After inoculating the microbial inoculants, the secondary community (bacteria of *Pseudomonadales*) in the compost can speed up the degradation rate of *Aspergillus* spp. for cellulose owing to it can degrade the organic acid. Therefore, the bacteria of *Pseudomonadales* make up a new predominant community with *Aspergillus* spp.

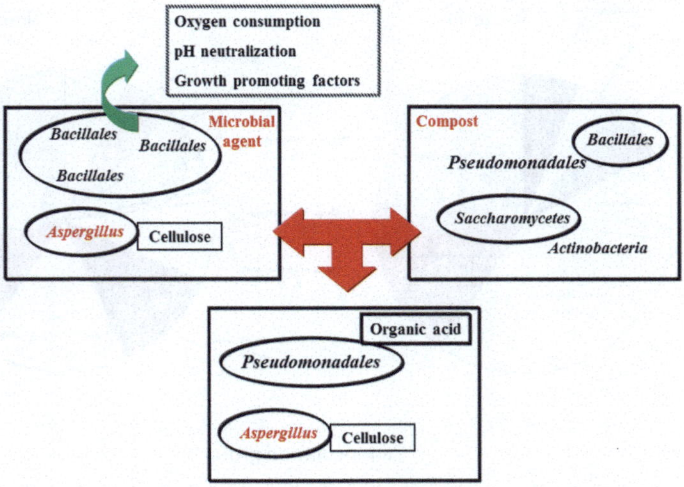

Fig. 3.5 The interaction of microbial communities in microbial agent and indigenous communities

3.3.2 Composing Technology for Rapid Humification of Kitchen Wastes

3.3.2.1 Experimental Design and Implementation

In the early stage of composting of kitchen wastes, material acidification can inhibit the activity of microorganisms to thus further result in long period and poor fertilizer efficiency of the composting of kitchen wastes. In order to solve the problem, the bio-augmented composting technology for oriented humification of kitchen wastes was explored by utilizing acidification-resistant complex microbial inoculants in order to realize composting for rapid humification of kitchen wastes. The kitchen wastes were manually mixed with bran, with the C/N ratio of the compost of 25.4 and the moisture content of 60.7%. Three experimental groups were set, including an inoculated group, an alkalified group and a control group. For the inoculated group, 1.25 mL of bacterial liquid of acidification-resistant complex microbial inoculants at exponential phase was added to per kilometer of dry raw materials of compost (concentration of microbial inoculants was 1×10^8 CFU mL^{-1}). For alkalified group, chemical buffers of 0.05 M MgO and 0.1 M K_2HPO_4 per kilometer of dry raw materials were added. The natural compost without the addition of microbial inoculants and buffers was taken as the control group. In the composting process, the ventilator capacity was controlled as 0.5 L min^{-1} kg^{-1} and also the change of temperature was monitored in real time. Samples were taken after the composting lasted for 0, 3, 8, 13, 19, 35 and 47 days, respectively. Before each sampling, the compost materials were manually mixed. The multiple samples collected at the different depths in pile centre were taken to make into a sample (about 200 g), which was stored at −20 °C for subsequent utilization.

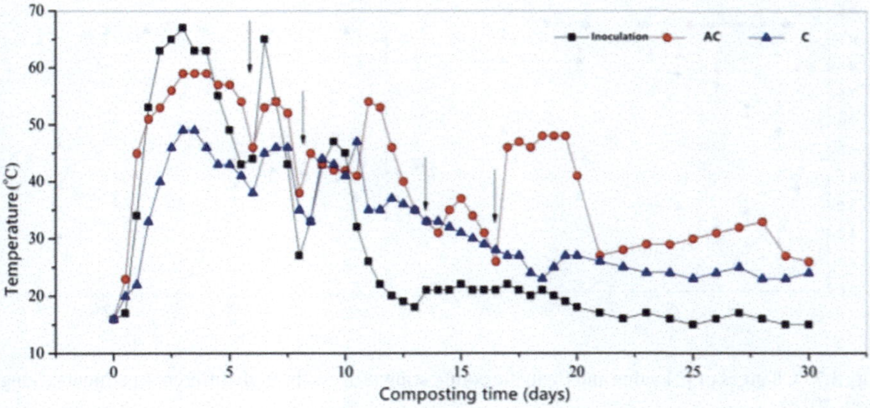

Fig. 3.6 Changes of temperature during different composting treatments. Inoculation–the inoculation with AAMC, AC–the alkaline compound treatment, C–the control group. Arrows represent the turning (Song et al. 2018)

It can be seen from Fig. 3.6 that the inoculated group had entered the high-temperature period (higher than 50 °C) 32 h after starting the composting, showing the highest temperature of 67 °C, followed by the alkalified group (59 °C). During the whole composting period, the temperature of the control group was lower than 50 °C. According to stipulation of the UK Environment Agency, the compost pile can conform to non-hazardous requirement as long as it is kept above 55 °C for 3–5 days. According to the stipulation, the inoculated group and the alkalified group can satisfy the non-hazardous requirement after they were composted for 5 and 6 days, respectively.

Due to the generation of a great number of micromolecular organic acids in the early stage, pH value significantly declined in the early stage of the composting. The pH of the inoculated group constantly rose during the whole composting period and finally stabilized. It can be seen that inoculating acidification-resistant complex microbial inoculants can avoid pH fluctuation of the composting system. With the generation of micromolecular organic acids in the early stage of composting, as well as the utilization of the micromolecular organic acids and generation of NH_3 in the later period, the pH value of the alkalified group decreased at first and then increased. The control group showed a significant acidification process: after composting for 3 days, the pH reduced to 4.86 and then gradually grew.

From the perspective of land utilization of organic fertilizers, the compost products with high conductivity show an unfavorable influence on soil microorganisms and the conductivity of compost products suitable for crop growth should be lower than 2.6 ms/cm. Figure 3.7b reveals that the conductivities of inoculated and alkalified groups both rose at first and then declined. In the early stage of composting, owing to macromolecular matters were mineralized into micromolecular compounds, the conductivity rose. With the advancement of composting process, the micromolecular compounds were gradually transformed into stable humus matters and therefore

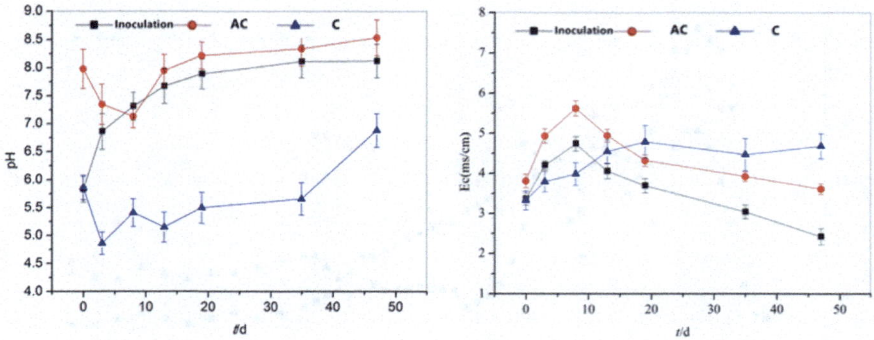

Fig. 3.7 Changes of pH value and Ec in the composting process through different treatments (Song et al. 2018)

conductivity gradually reduced. The conductivity of the control group constantly rose. Except for the inoculated group, the conductivities of the other two treatment groups were both higher than 2.6 ms/cm.

3.3.2.2 Composting Mechanism for Rapid Humification by Adding Acidification-Resistant Complex Microbial Inoculants

Acetic acid mainly came from pyruvic acid, which generated acetic acid through glycolysis based on pyruvate dehydrogenase and aldehyde dehydrogenase. By using acetyl-coenzyme A synthetase, acetic acid produced acetyl coenzyme A, which was finally subjected to tricarboxylic acid (TCA) cycle and further utilised through glyoxylate cycle branch. The biochemical reactions triggered by acetyl-coenzyme A synthetase and malate synthase both appeared in the direction of degradation of acetic acid. Compared with the control group, the diversities of microorganisms generating acetyl-coenzyme A synthetase and malate synthase in the inoculated group both greatly grew. It revealed that inoculating acidification-resistant complex microbial inoculants can significantly improve four key enzymes (A, B, C and D) in the metabolic pathways of acetic acid. Enzyme A denotes pyruvate dehydrogenase, which was generated from *Staphylococcus* spp., *Pseudomonas* spp. and *Bacillus* spp. in the inoculated group. In the alkalified group, enzyme A was generated by *Escherichia* spp., *Pseudomonas* spp., *Bacillus* spp. and *Mycobacterium* spp., while it was produced by *Mycobacterium* spp. in the CK group. Enzyme B refers to aldehyde dehydrogenase, which was generated by *Rhodococcus* spp. both in inoculated and alkalified groups while no enzyme B was detected in the CK group. In the metabolic pathways of acetic acid, the biochemical reactions facilitated by enzymes A and B both occurred at the side where acetic acid was generated. Compared with the control group, the diversities of microorganisms generating enzymes A and B both rose in inoculated and alkalified groups. However, compared with alkalified group, the diversity of microorganisms generating enzymes A and B in the inoculated group was

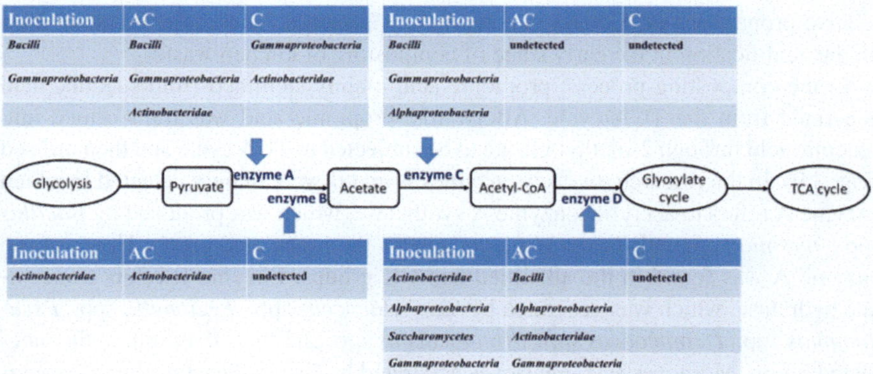

Fig. 3.8 Acetate metabolic pathway, enzyme A–Pyruvate dehydrogenase, enzyme B–Aldehyde dehydrogenase, enzyme C–Acetyl-coenzyme A synthetase, enzyme D–Malate synthase, Inoculation–the inoculation with AAMC, AC–the alkaline compound treatment, C–the control group (Song et al. 2018)

insignificant. Enzyme C denotes acetyl-coenzyme A synthetase. In the inoculated group, enzyme C was generated by *Bacillus* spp., *Yersinia* spp. and *Bradyrhizobium* spp. while no enzyme C was detected in the alkalified and CK groups. Enzyme D refers to malate synthase. In the inoculated group, enzyme D was produced by *Mycobacterium* spp., *Bradyrhizobium* spp., *Saccharomyces* spp., *Escherichia* spp., *Rhodococcus* spp., *Pseudomonas* spp. and *Candida* spp. while it was generated by *Geobacillus* spp., *Rhodopseudomonas* spp., *Rhodococcus* spp. and *Pseudomonas* spp. in the alkalified group. However, enzyme D was not detected in the CK group.

Figure 3.8 quantitatively displays the proportions of the four key enzymes in metabolic pathways in metabolism of carbohydrates under different composting treatments. It can be seen from the Figure that in terms of proportion of enzyme A, the three composting treatments were displayed in a ascending order as inoculated group, alkalified group and CK group. Moreover, the proportion (0.55%) of enzyme B in the inoculated group was also significantly lower than that (1.40%) in the alkalified group. Although the diversities of bacteria generating enzymes A and B in the inoculated group were improved, the proportions of enzymes A and B were greatly lower than those in the other two groups. It probably explained why the improvement of diversity of bacteria generating these enzymes in the acetic acid formation direction of the aforementioned biochemical reaction failed to cause the great accumulation of acetic acid. The enzymes C and D in the inoculated group showed the largest proportions (2.21% and 4.42%, respectively), followed by the alkalified group (0% and 4.20%). The two enzymes were not found in the CK group. It was also the possible reason why the improvement of diversity of bacteria generating these enzymes in the acetic acid formation direction of the aforementioned biochemical reaction did not cause the great accumulation of acetic acid. Inoculating acidification-resistant complex microbial inoculants not only promoted the catalysis towards the direction of acetic acid degradation but also can improve the diversity and relative proportions of bacteria generating key enzymes. Thus, it avoids inhibition on the acidification in the early stage of composting of kitchen wastes.

In the composting process, propionic acid mainly stemmed from succinic acid generated from the TCA cycle. Afterwards, propionic acid was transformed into succinic acid through 2-methyl-citrate to be subjected to TCA cycle and then utilised (Fig. 3.9). In the research, two key enzymes A and B were mainly detected in which enzyme A refers to acetyl-coenzyme A synthetase, which was produced by *Bacillus* spp., *Yersinia* spp. and *Bradyrhizobium* spp. in the inoculated group. However, no enzyme A was found in the alkalified and CK groups. Enzyme B refers to aconitate hydratase, which was generated by *Staphylococcus* spp., *Legionella* spp., *Pseudomonas* spp., *Deinococcus* spp., *Streptococcus* spp. and *Rickettsia* spp. in the inoculated group. Moreover, enzyme B was generated by *Legionella* spp., *Pseudomonas* spp., *Escherichia* spp., *Corynebacterium* spp., *Rickettsia* spp. and *Mycobacterium* spp. in the alkalified group while it only came from *Pseudomonas* spp. in the CK group. The biochemical reactions under the catalytic effect of enzymes A and B both appeared at the propionic degradation direction. Compared with the control group, the diversity of microorganisms generating enzyme A in the inoculated group A greatly rose while that in the alkalified group insignificantly changed. The diversities of microorganisms producing enzyme B all significantly increased. Inoculating acidification-resistant complex microbial inoculants can greatly improve the diversity of strains generating the two key enzymes in metabolic pathways of propionic acid.

The proportion (2.21%) of enzyme A in the inoculated group was significantly larger than those in the other two groups. In terms of proportion of enzyme B, the three composting treatments were displayed in a descending order as alkalified group (7.69%), inoculated group (5.52%) and CK group (1.89%). Inoculating acidification-resistant complex microbial inoculants and adding chemical buffers both can increase the proportion of enzymes which can catalyse the biochemical reaction at the propionic acid degradation direction.

3.3.3 Composting Technology for Humification of Mixed Materials

3.3.3.1 Experimental Design and Implementation

For the urgent demand for resourceful utilization of agricultural wastes, the composting technology for oriented humification of mixed materials of biogas residue and excrements of livestocks (pig manure and chicken manure) was explored (Table 3.4). On this basis, the technical control of mixed composts with different proportions, characteristics of DOM spectroscopy and evolutionary change law of diversity of microorganisms were separately analyzed. In this way, the optimal technological parameters of composting for oriented humification of agricultural mixed materials

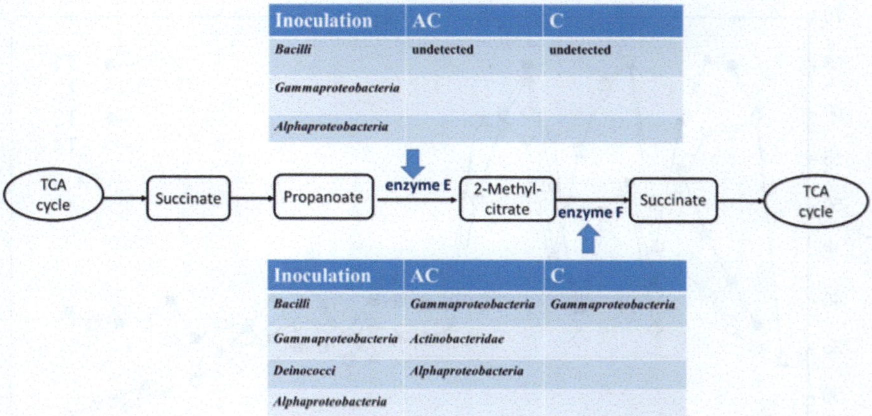

Fig. 3.9 Propanoate metabolism pathway, enzyme E–Acetyl-coenzyme A synthetase, enzyme F–Aconitate hydratase, Inoculation–the inoculation with AAMC, AC–the alkaline compound treatment, C–the control group (Song et al. 2018)

Table 3.4 Proportions of mixed materials

Factors	Renewal (kg)	Pig manure (kg)	Chicken manure (kg)
T1	7.02	7.425	8.19
T2	5.85	7.425	6.825
T3	5.85	8.49	8.19
T4	7.02	8.49	6.825

were obtained. Four experimental groups (T1–T4) were designed by using orthogonal test. The mixed materials of the compost with the initial moisture content of 60% showed pH value in the range of 6.5–7.5 and C/N ratio ranged from 20 to 25. Moreover, the ventilatory capacity was controlled at 0.5 L/(kg h) and the particles of materials exhibited the diameter in the range of 1.5–5 cm. After composting for 30 days, same quantities of samples were separately taken from the points at the heights of 15, 20 and 25 cm in the centre of the compost and then uniformly mixed, frozen-dried, ground and screened for subsequent utilization.

In the early stage of composting, the heating rate and high-temperature (higher than 50 °C) duration of T2 and T3 were greatly larger than those of T1 and T4, all showing the highest temperature of 65.5 °C. The high-temperature durations of T2 and T3 were 132 and 120 h, respectively, reaching the harmless level according to sanitary standard for the harmless treatment of excrements. However, T1 and T4 did not satisfy the standard. Additionally, T3 showed the largest heating rate and the longest high-temperature duration and maximum highest temperature at the high-temperature period, followed by T2, and T1 and T4 exhibited the lowest level (Fig. 3.10).

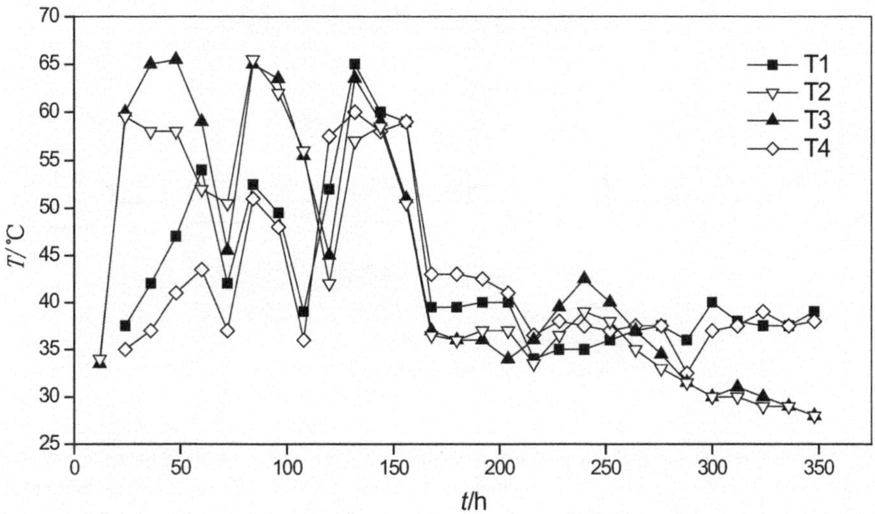

Fig. 3.10 Changes of temperature during composting

The pH values of different experimental groups all rose at first, then declined and finally increased. The pH value steadily varied and fluctuated at around 8.0, and finally reached the alkalescent standard of mature compost. At the high-temperature period of composting, organics were rapidly decomposed to further generate micro-molecular organic acids, which led to the reduction of pH value. At the end of composting, micromolecular organic acids were consumed out and the remained macromolecular humic substances caused the slight growth of pH value, so that the system appeared as alkalescence (Fig. 3.11).

Some scholars suggested that a high content of NH_4^+-N represents instability of compost and put forward that the NH_4^+-N content of compost products of municipal wastes should be lower than 0.04%. After composting for 30 days, in terms of NH_4^+-N content, the composting treatments were displayed in an ascending order as T3, T2, T1 and T4, in which the NH_4^+-N contents of T3, T2 and T1 were 0.25 mg/g, 0.34 mg/g and 0.35 mg/g, respectively, satisfying the standard of maturation of compost. By contrast, T4 failed to reach the maturation standard in terms of NH_4^+-N content (Fig. 3.12).

3.3.3.2 Evolutionary Change Law of Bacterial Communities During the Composting of Mixed Materials

The biodiversity index is an important indicator for effectively characterizing richness and distribution uniformity of species in ecological environment. By counting the number of bands in DGGE atlas and calculating Shannon-Weaver index, the bacterial diversity during composting was analyzed (Fig. 3.13). The number of

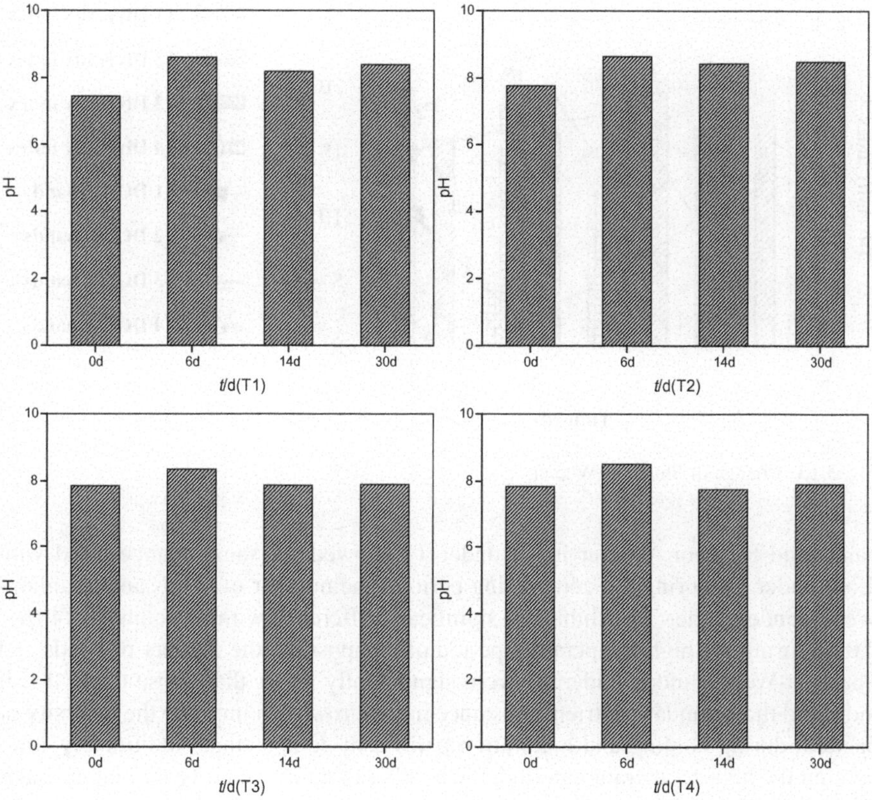

Fig. 3.11 Changes of pH value of different treatments during composting

Fig. 3.12 Changes of NH_4^+-N of different treatments during composting

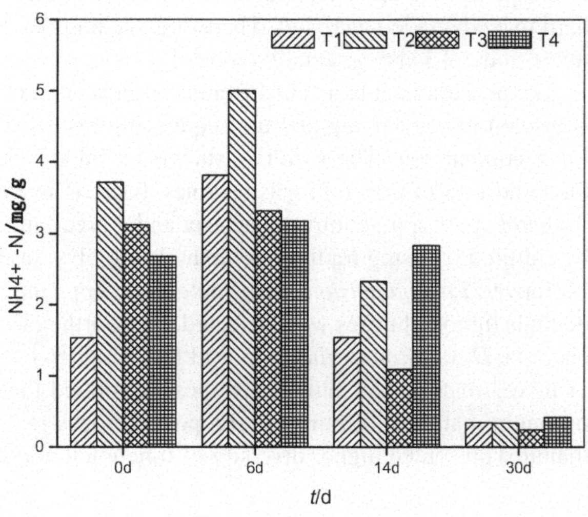

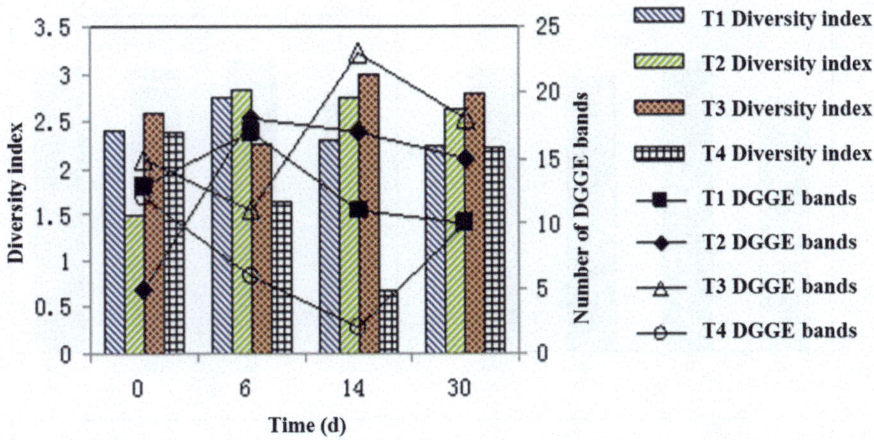

Fig. 3.13 Change of Shannon-Winner

bands and Shannon-Weaver index under T1 showed the same change trend with those under T2 during the composting period. The number of bands and Shannon-Weaver index under T3 exhibited a significant difference with those under T4 ($p < 0.05$): during the high-temperature period of composting, the number of bands and Shannon-Weaver index under T3 were significantly larger than those under T4. It indicated that abundant nutrient substance in windrows can improve the diversity of bacteria during cooling and maturation periods of composting. Additionally, after passing the high-temperature period, the number of bands and diversity index under T3 greatly declined and then they significantly rose owing to the temperature of the compost decreased. In the end of composting, the organic matters in windrows prone to be degraded were consumed out and the wood-fibre macromolecular substances hard to be degraded were left. Therefore, the number of bands and Shannon-Weaver index under T3 also gradually reduced.

The predominant bands in different experimental groups during composting were subjected to sequencing and the sequencing results were deposited into Genebank for a comparison. The result is shown in Table 3.5. During the humification of the windrows of mixed biogas residues, bacteria in *Proteobacteria* and *Firmicutes*, *Acinetobacter* spp., compost bacteria and a great number of bacteria which cannot be cultured by using traditional methods but play an important role, were detected. Moreover, *Fusobacterium* spp., *Bacteroides* spp. and *Pseudomonas* spp. which can degrade lignocelluloses were detected. It is worth noting that non-cultured anaerobic bacteria, *Desulfotomaculum* spp. and *Psychrobacter* spp. were found in the compost of mixed materials of biogas residues. It implied that compared with the compost of single materials, the mixed composts (biogas residue, pig manure and chicken manure) presented higher diversity of functional microorganism.

Table 3.5 Blast results of the band sequences

Serial number	Accession number	The species with the greatest similarity	Similarity (%)
A	HQ731452.1	*Acinetobacter* sp. BN-HKY5	100
B	JQ069959.1	*Psychrobacter* sp. BSw21684	99
C	EU551120.1	Uncultured *Bacteroidetes* bacterium clone B9	100
D	DQ141183.1	*Ruminofilibacter xylanolyticum* strain S1	100
E	FJ599513.1	*Clostridium thermocellum* strain CTL-6	98
F	JQ269284.1	Bacterium WHC3-10	100
G	JN834841.1	Uncultured *Bacteroidetes* bacterium clone 181T2TEAT	99
H	FN436048.1	Uncultured bacterium clone HAW-R60-B-727d-F	99
I	HQ433472.2	*Pseudomonas* sp. A84(2010)	100
J	AB436739.1	*Desulfonosporus* sp. AAN04	95
K	DQ839147.1	*Clostridium* sp. enrichment clone Lace	95
L	JQ268616.1	Uncultured anaerobic bacterium clone XA3	100
M	JF915345.1	*Acinetobacter junii* strain NW123	100
N	HQ731452.1	*Acinetobacter* sp. BN-HKY5	98
O	EF629964.1	Uncultured *alpha proteobacterium* clone 6mML2F12	100
P	FJ675561.1	Uncultured bacterium clone LL141-8K15	100
Q	CU918603.1	Uncultured *Firmicutes* bacterium clone QEEA1AB12	93
R	AB437998.1	Uncultured compost bacterium	97
S	FN667411.1	Uncultured compost bacterium clone PS2677	100
T	AM491470.1	*Psychrobacter* sp. Nj-69 16S rRNA gene, strain Nj-69	100
U	AB507775.1	Uncultured *Firmicutes* bacterium	96
V	EF414143.1	Uncultured *alpha proteobacterium* clone MPWIC_A11	100
W	HQ224821.1	Uncultured bacterium clone ABRB33	100
X	DQ141183.1	*Ruminofilibacter xylanolyticum* strain S1	100
Y	JN368260.1	Uncultured bacterium clone LSW-L1-126	100
Z	JN834841.1	Uncultured *Bacteroidetes* bacterium clone 181T2TEAT	100
A’	HQ154860.1	Uncultured prokaryote clone 08061601-ZSS_YX_Z8_EU_2_46	100

3.3.4 Control Technology for Odorous Pollutants During Composting

3.3.4.1 Emission Law of Odorous Pollutants During Mechanical Separation and Biological Pre-treatment

Conducting mechanical separation based on the difference of physical properties of different components of municipal solid wastes is a key technological link of quality-based resourceful utilization of complex wastes. The dispersion of odor pollutants caused by various mechanical actions (including turning, winnowing and shaking) during the separation is an important secondary pollution source. Based on limited emission concentrations of eight odor pollutants specified in standard GB14554-93 (as shown in Table 3.6), the emission law of odor pollutants during technological links (the mechanical separation and biological pre-treatment) of municipal solid wastes was analyzed.

The results showed that the concentration of methyl mercaptan in fresh wastes was far larger than the national standard in which the methyl sulfide showed the lowest concentration (0.0236 mg/m^3) while the concentration (0.395 mg/m^3) of styrene was the largest. During mechanical separation, the concentration of odorous gases generated by municipal wastes gradually declined. However, during the screening link using a rotary screen, the odor concentrations of ammonia, hydrogen sulfide, carbon disulfide and methyl sulfide slightly rose. Moreover, the odor concentrations of ammonia, hydrogen sulfide, carbon disulfide, styrene and methyl sulfide at rotary screen were 11.67%, 3.8%, 25.42%, −17.67% and 128.93% higher compared with those at the manual sorting table, respectively.

Table 3.6 Standard values of odor pollutants from factory sector

Serial number	Control item	Unit	First grade	Second grade		Third grade	
1	Ammonia	mg/m^3	1	1.5	2	4	5
2	Trimethylamine	mg/m^3	0.05	0.08	0.15	0.45	0.8
3	Hydrogen sulfide	mg/m^3	0.03	0.06	0.1	0.32	0.6
4	Methyl mercaptan	mg/m^3	0.004	0.007	0.01	0.02	0.035
5	Methyl sulfide	mg/m^3	0.03	0.07	0.15	0.55	1.1
6	Dimethyl disulfide	mg/m^3	0.03	0.06	0.13	0.42	0.71
7	Carbon disulfide	mg/m^3	2	3	5	8	10
8	Styrene	mg/m^3	3	5	7	14	19
9	Odor concentration	Dimensionless	10	20	30	60	70

For biological treatment process, the NH_3 concentration increased at first and then reduced with time of biological treatment, and it reached the maximum at the second day. This was because NH_4^+ was transformed into liquid NH_3 through deprotonation and further transformed into gaseous NH_3, which was further diffused into atmosphere. After the biological treatment was ended, the NH_3 concentration accounted for about 1/3 of that in the early stage of composting. Moreover, the H_2S concentration also rose at first and then reduced with biological treatment time, reaching the maximum (0.252 mg/m^3) at the fourth day. It was likely related to that in the early stage of composting, microbial metabolism consumed lots of oxygen, causing the anaerobic environment to appear at the local area of the pile. The CS_2 concentration reached the maximum (0.281 mg/m^3) at the fourth day of the biological treatment, which was generally higher than concentrations of the other pollutants. Moreover, the threshold value of CS_2 is low and therefore people can suffer from acute and chronic poisoning and even retinopathy when sucking a certain dosage. Therefore, during the biological treatment, the CS_2 should be considered as an odor pollutant to be controlled preferentially. The concentrations of styrene, methyl mercaptan and methyl sulfide were low and gradually decreased with the conduction of biological treatment.

3.3.4.2 Emission Law of Volatile Organic Compounds During Mechanical Separation and Biological Pre-treatment

The main pollutants in odorous gas are highly related to the compositions of municipal wastes. The compositions of volatile organic compounds (VOCs) in primary wastes were analyzed by applying gas chromatography-mass spectrometry (GC-MS). As shown in Table 3.7, the VOC composition of primary wastes was complex, mainly shown as hydrocarbons and benzene series, which involved 12 hydrocarbon compounds, 9 aromatic compounds, one sulfide, one ketone compound and an ester compound. The concentration of total VOCs was 8.4606 mg/m^3.

The results of compositions of odorous gas of municipal wastes during mechanical pre-treatment showed that compared with primary wastes, during primary separation, the concentrations of various benzene series and ketones (including benzene, chlorobenzene, trichlorotoluene, m-dichlorobenzene, p-dichlorobenzene, dichloromethane, tetrachloroethylene, methylbenzene, trimethylbenzene, 1-ethyl-4-methylbenzene, ethyl acetate, n-heptane, n-hexane, methyl ethyl ketone, methyl isobutyl ketone and methyl-n-butyl ketone) generally rose (Table 3.8). Especially, the concentration of dichloromethane was 6.35 times that of primary wastes, which was likely related to that mechanical motion during separation speeded up the dispersion of gases. In industry, dichloromethane shows a great irritation on skin and mucosa and therefore dichloromethane should be intensively controlled during mechanical pre-treatment. The concentrations of olefin pollutants generally declined and especially cis-1,3-dichloropropene, hexachlor-1,3-butadiene, trans-1,3-dichloropropene were not detected during mechanical pre-treatment.

Table 3.7 Concentrations of VOCs during MSW

Compound	Detected concentration (mg/m^3)	Boiling point (°C)	Compound	Detected concentration (mg/m^3)	Boiling point (°C)
Benzene	0.0623	80.1	Trichloro ethylene	0.2020	86.7
Methyl bromide	0.0005 L	3.6	1,2,4-trichlorobenzene	0.0005 L	214
Carbon tetrachloride	0.0005 L	76.8	Trimethylbenzene	0.1530	164.7
Chlorobenzene	0.0106	131.7	Chloroethylene	0.0005 L	−13.9
Trichloromethane	0.0149	61	Xylene	0.6193	144.4
Chloromethane	0.0005 L	−24	Acetone	0.0005 L	56.48
Cis-1,3-dichloropropene	0.0155	104	Bromodichloromethane	0.0005 L	90.1
Trans-1,2-dichloroethylene	0.0005 L	47.5	1,3-butadiene	0.0005 L	−4.5
1,2-dibromoethane	0.0005 L	131.4	Carbon disulfide	0.1064	46–47
o-dichlorobenzene	0.0005 L	180.4	Chloromethylbenzene	0.0005 L	–
m-dichlorobenzene	0.0292	139.1	3-chlorine-1-propylene	0.0005 L	44.6
p-dichlorobenzene	0.0299	138.35	Cyclohexane	0.0005 L	80.74
1,1-dichloroethane	0.0121	57.3	Bromodicloromethane	0.0005 L	119–120
1,2-dichloroethane	0.0005 L	83.5	1,4-dioxane	0.0005 L	101.4
1,1-dichloroethane	0.0005 L	32	1-ethyl-4-methylbenzene	0.0921	162
1,2-propylene dichloride	0.3087	96.8	Ethyl acetate	1.5017	77.2
Ethylbenzene	0.1644	136.19	Trans-1,2-dichloroethylene	0.0005 L	47.5
Chloroethane	0.0005 L	12.5	n-heptane	0.0934	98.5
Trichlorofluoromethane	0.0005 L	23.7	n-hexane	0.0383	68.7
Trichlorine trifluoromethane	0.0005 L	–	Isooctane	0.0005 L	99.2

(continued)

Table 3.7 (continued)

Compound	Detected concentration (mg/m^3)	Boiling point (°C)	Compound	Detected concentration (mg/m^3)	Boiling point (°C)
Dichloride tetrafluoromethane	0.0005 L	–	Isopropanol	0.0005 L	82.5
Dichlorodifluoromethane	0.0005 L	−29.8	Methyl ethyl ketone	0.8748	80
Hexachlor-1,3-butadiene	0.0014	210–215	Methyl isobutyl ketone	0.0005 L	115.8
Dichloromethane	1.1229	39.8	Methyl-n-butyl ketone	0.0005 L	128
Styrene	0.5007	145.2	Propylene	0.0005 L	−47.4
1,1,2,2-tetrachloroethane	0.0005 L	146.2	Methyl tert-butyl ether	0.0005 L	55.3
Tetrachloroethylene	0.0253	121.2	Tetrahydrofuran	0.0005 L	63.4
Methylbenzene	2.4693	110.63	Tribromethane	0.0005 L	146–151
Trans-1,3-dichloropropene	0.0125	104.3	Acetic acid vinyl ester	0.0005 L	71.8
Trichloroethane	0.0005 L	113.5	Bromethylene	0.0005 L	15.8
1,1,2-trichloroethane	0.0005 L	114	Total VOCS	8.4606	

Table 3.8 Concentrations of VOCs during mechanical pre-treatment

Compound	Detected concentration (mg/m^3)	Boiling point (°C)	Compound	Detected concentration (mg/m^3)	Boiling point (°C)
Benzene	0.1548	80.1	Trichloro ethylene	0.1644	86.7
Methyl bromide	0.0005 L	3.6	1,2,4-trichlorobenzene	0.0005 L	214
Carbon tetrachloride	0.0005 L	76.8	Trimethylbenzene	0.2628	164.7
Chlorobenzene	0.0171	131.7	Chloroethylene	0.0005 L	−13.9
Trichloromethane	0.0155	61	Xylene	0.5934	144.4
Chloromethane	0.0005 L	−24	Acetone	0.0005 L	56.48
Cis-1,3-dichloropropene	0.0005 L	104	Bromodichloromethane	0.0005 L	90.1
Trans-1,2-dichloroethylene	0.0005 L	47.5	1,3-butadiene	0.0005 L	−4.5
1,2-dibromoethane	0.0005 L	131.4	Carbon disulfide	0.1019	46–47
o-dichlorobenzene	0.0005 L	180.4	Chloromethylbenzene	0.0005 L	–
m-dichlorobenzene	0.0902	139.1	3-chlorine-1-propylene	0.0005 L	44.6
p-dichlorobenzene	0.0926	138.35	Cyclohexane	0.0005 L	80.74
1,1-dichloroethane	0.0199	57.3	Chlorodibromomethane	0.0005 L	119–120
1,2-dichloroethane	0.0005 L	83.5	1,4-dioxane	0.0005 L	101.4
1,1-dichloroethylene	0.0005 L	32	1-ethyl-4-methylbenzene	0.1644	162
1,2-propylene dichloride	0.2585	96.8	Ethyl acetate	3.0800	77.2
Ethylbenzene	0.1570	136.19	Trans-1,2-dichloroethylene Dichloroethylene	0.0005 L	47.5
Chloroethane	0.0005 L	12.5	n-heptane	0.4187	98.5
Trichlorofluoromethane	0.0005 L	23.7	n-hexane	1.4003	68.7

(continued)

Table 3.8 (continued)

Compound	Detected concentration (mg/m^3)	Boiling point (°C)	Compound	Detected concentration (mg/m^3)	Boiling point (°C)
Trichlorine trifluoromethane	0.0005 L	–	Isooctane	0.0005 L	99.2
Dichloride tetrafluoromethane	0.0005 L	–	Isopropanol	0.0005 L	82.5
Dichlorodifluoromethane	0.0005 L	−29.8	Methyl ethyl ketone	2.1088	80
Hexachlor-1,3-butadiene	0.0005 L	210–215	Methyl isobutyl ketone	0.6144	115.8
Dichloromethane	7.1729	39.8	Methyl-n-butyl ketone	0.0506	128
Styrene	0.4704	145.2	Propylene	0.0005 L	−47.4
1,1,2,2-tetrachloroethane	0.0005 L	146.2	Methyl tert-butyl ether	0.0005 L	55.3
Tetrachloroethylene	0.0498	121.2	Tetrahydrofuran	0.0005 L	63.4
Methylbenzene	5.3951	110.63	Tribromethane	0.0005 L	146–151
Trans-1,3-dichloropropene	0.0005 L	104.3	Acetic acid vinyl ester	0.0005 L	71.8
Trichloroethane	0.0005 L	113.5	Bromethylene	0.0005 L	15.8
1,1,2-trichloroethane	0.0005 L	114	Total VOCs	22.9553	

During the biological pre-treatment, the concentrations of VOCs generally declined due to the utilization of microorganisms. As shown in Table 3.9, the detected concentration of the total VOCs was 3.8091 mg/m^3, which was 2.2 times lower than that of primary wastes. Among 64 VOCs defined by using TO-14A and TO-15 methods of environmental protection agency (EPA) of the United States, 38 VOCs were not detected. At the end of pre-treatment, ketone substances were found, which indicated that alcohol, ketone, etc. were all generated due to the anaerobic reaction occurring in local areas in the composting process.

3.3.4.3 Evaluation on the Treatment Effect of Deodorizing Technology Using a Biofilter

By using a biofilter, the odorous gas generated in treatment plants of municipal wastes during mechanical separation, biological pre-treatment and composting were collected and removed. The change of concentration of pollutants before and after removing odorous gas is shown in Table 3.10. The result showed that the biofilter can favorably remove styrene and methyl sulfide, with the removal rate of 100%. Moreover, the removal rates of NH_3 and H_2S by using the biofilter were 38.10% and 24.05%, respectively, and the removal rate (17.22%) of methyl mercaptan was the lowest. It was probably that the temperature in summer rose and therefore gas evaporated to thus take lots of moisture, further leading to the reduction of moisture content of fillers. As a result, part of fillers is hardened to thus result in a low removal rate of odorous gas. Therefore, it is necessary to increase moisture in filter materials and it is necessary to maintain the moisture content of filter materials in the range of 55–65%.

3.4 Evaluation on Utilisability of Compost Products

3.4.1 Evaluation on Utilisability of Compost Products of Kitchen Wastes

3.4.1.1 Germination Index

The toxicity test of plants is an effective and common method for evaluating the maturity of compost, which mainly includes germination test, growth test and enzyme activity, in which the seed germination experiment shows significant superiority. Researchers generally suggested that if the germination index (GI) is in the range of 80–85%, the compost has been mature or shows no phytotoxicity, having a positive effect on plant growth. By taking composting technology for rapid humification of kitchen wastes as an example, after the composting lasted for 8 days, the GI of the inoculated group was larger than 80% while that of the alkalified group was lower

Table 3.9 Concentrations of VOCs during biological pre-treatment

Compound	Detected concentration (mg/m^3)	Boiling point (°C)	Compound	Detected concentration (mg/m^3)	Boiling point (°C)
Benzene	0.0192	80.1	Trichloro ethylene	0.0287	86.7
Methyl bromide	0.0005 L	3.6	1,2,4-trichlorobenzene	0.0005 L	214
Carbon tetrachloride	0.0005 L	76.8	Trimethylbenzene	0.1559	164.7
Chlorobenzene	0.0074	131.7	Chloroethylene	0.0005 L	−13.9
Trichloromethane	0.0051	61	Xylene	0.3282	144.4
Chloromethane	0.0005 L	−24	Acetone	0.0274	56.48
Cis-1,3-dichloropropene	0.0005 L	104	Bromodichloromethane	0.0005 L	90.1
Trans-1,2-dichloroethylene	0.0005 L	47.5	1,3-butadiene	0.0005 L	−4.5
1,2-dibromoethane	0.0005 L	131.4	Carbon disulfide	0.1444	46–47
o-dichlorobenzene	0.0005 L	180.4	Chloromethylbenzene	0.0005 L	–
m-dichlorobenzene	0.0745	139.1	3-chlorine-1-propylene	0.0005 L	44.6
p-dichlorobenzene	0.0765	138.35	Cyclohexane	0.0209	80.74
1,1-dichloroethane	0.0022	57.3	Chlorodibromomethane	0.0005 L	119–120
1,2-dichloroethane	0.0005 L	83.5	1,4-dioxane	0.0005 L	101.4
1,1-dichloroethylene	0.0005 L	32	1-ethyl-4-methylbenzene	0.1441	162
1,2-propylene dichloride	0.0262	96.8	Ethyl acetate	0.0005 L	77.2
Ethylbenzene	0.4128	136.19	Trans-1,2-dichloroethylene	0.0005 L	47.5
Chloroethane	0.0005 L	12.5	n-heptane	0.0564	98.5
Trichlorofluoromethane	0.0005 L	23.7	n-hexane	0.0737	68.7
Trichlorine trifluoromethane	0.0005 L	–	Isooctane	0.0005 L	99.2

(continued)

Table 3.9 (continued)

Compound	Detected concentration (mg/m^3)	Boiling point (°C)	Compound	Detected concentration (mg/m^3)	Boiling point (°C)
Dichloride tetrafluoromethane	0.0005 L	–	Isopropanol	0.0005 L	82.5
Dichlorodifluoromethane	0.0005 L	−29.8	Methyl ethyl ketone	0.5002	80
Hexachlor-1,3-butadiene	0.0032	210–215	Methyl isobutyl ketone	0.0005L	115.8
Dichloromethane	0.6498	39.8	Methyl-n-butyl ketone	0.0315	128
Styrene	0.0874	145.2	Propylene	0.0005 L	−47.4
1,1,2,2-tetrachloroethane	0.0005 L	146.2	Methyl tert-butyl ether	0.0005 L	55.3
Tetrachloroethylene	0.0352	121.2	Tetrahydrofuran	0.0005 L	63.4
Methylbenzene	0.9252	110.63	Tribromethane	0.0005 L	146–151
Trans-1,3-dichloropropene	0.0005 L	104.3	Acetic acid vinyl ester	0.0005 L	71.8
Tichloroethane	0.0005 L	113.5	Bromethylene	0.0005 L	15.8
1,1,2-trichloroethane	0.0005 L	114	Total VOCs	3.8091	

Table 3.10 Analysis of biofilter removal rate

Gas	Initial concentration (mg/m^3)	Outgassing concentration (mg/m^3)	Outgassing rate (kg/h)	Removal rate (%)
NH_3	0.042	0.026	1.26×10^{-5}	38.10
H_2S	0.0158	0.012	8.85×10^{-5}	24.05
CS_2	0.727	0.2804	3.53×10^{-4}	61.43
Styrene	0.3178	0	–	100
Methyl mercaptan	0.00906	0.0075	0.00906	17.22
Methyl sulfide	0.01546	0	–	100%

Note Dimethyl disulfide and trimethylamine were not detected during mechanical separation and biological pre-treatment, Therefore, the odorous gas removed by using the biofilter did not include dimethyl disulfide and trimethylamine

than 10%. After the composting lasted for 13 days, the GI of the inoculated group was up to 111% while that of alkalified group was about 40%. It can be seen that the inoculated group satisfied the harmless standard after being composted for 5 days and mature after being composted for 8 days. The aforementioned result indicated that inoculating acidification-resistant complex microbial inoculants can greatly speed up the humification process, improve the composting efficiency of kitchen wastes and shorten the composting period to 8 days (Fig. 3.14).

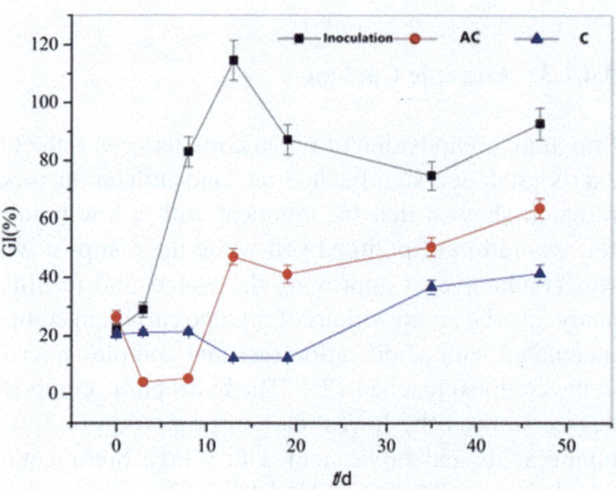

Fig. 3.14 Changes of germination index during different composting

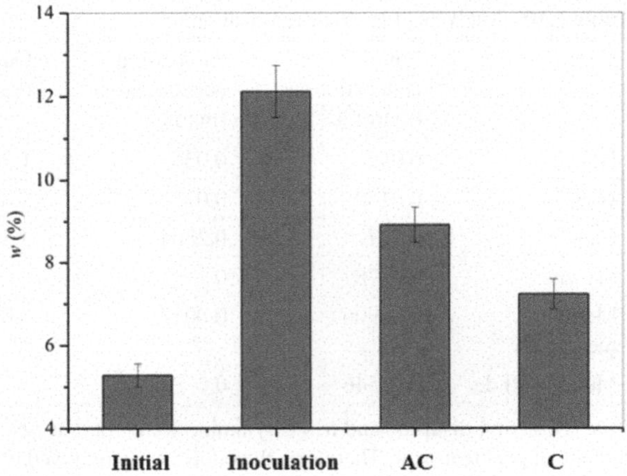

Fig. 3.15 Changes of HULIS-C during different composting treatments

3.4.1.2 Carbon in Humic-Like Substances

Compared with the control group, the carbon in humic-like substances (HULIS-C) in the compost in the inoculated group improved by 67.63% while that of the alkalified group increased by 23.24%. This implies that inoculating complex microbial system for degrading organic acids can significantly increase the humic content of the compost (Fig. 3.15). By comparing with the compost in the control group, it can be seen that chemical buffers also can promote the synthesis of humus of kitchen waste compost to some extent.

3.4.1.3 Organic Carbon

Structural composition of organic matters can reflect the quality of compost, which exerts guidance significance on land utilization mode of compost products. The research showed that the compost with a low humification degree was favorable for restoration of polluted soil while the compost with a high humification degree was conductive to improving the water- and fertility-retention ability of soil. By analyzing the compositions of organic carbon in compost products of kitchen wastes inoculated with acidification-resistant complex microbial inoculants, the HULIS-C in the compost reached 12%. The hydrophilic compositions with a low humification degree showed the lowest carbon proportion (1.42%) while the sum of carbon in humic acids and fulvic acid with a high humification degree was up to 10.31%, which took up 85.07% of HULIS-C. The compost products of kitchen wastes were in favor of improving water- and fertility-retention ability of soil.

3.4.1.4 Evaluation of Utilisability of Compost Products

The conductivity of compost in the inoculated group was 2.26 ms/cm, satisfying the standard, and the compost contained 55% of organic matters, larger than the specified content (not lower than 45%) of organic matters in organic fertilizer (standard NY 525-2012). Moreover, the content of HULIS-C was up to 12%, which was remarkably higher than the requirement (not lower than 3.0%) on the content of humic acid in water-soluble fertilizers containing humic acid (standard NY1106-2010). Moreover, the GI was 114.6%, which was significantly larger than the value (80%) recommended by scholars. Overall, the compost products inoculated with acidification-resistant complex microbial inoculants show a high utilisability of plants, with a promising application prospect (Fig. 3.16).

3.4.2 Evaluation on Utilizability of Compost of Mixed Materials

3.4.2.1 GI

Under different treatments, GI constantly rose during the composting and reached 80–85% after the composting was ended, based on which the compost can be considered to be mature (Fig. 3.17). With the conduction of composting, the materials inhibiting plant growth gradually reduced, which was the reason why GI constantly increased. Under different treatments, the values of GIs showed a great difference at the same period of composting and the values of GIs under different treatments were greatly different at the early stage of composting. Different contents of materials (including volatile fatty acids and phenolic acids) capable of triggering phytotoxicity were produced under various material ratios, leading to a great difference in the values of GIs. After the composting was ended, the treatments were displayed in a descending order as T2, T3, T1 and T4 in terms of the values of GI.

3.4.2.2 Fecal Coliforms

The value of fecal coliforms in compost was detected in order to directly judge the harmless level of the compost. According to non-hazardous standard of faeces (GB 7959-87), the value of fecal coliforms of the compost being subjected to harmless treatment should reach the sanitary standard in the range of 10^{-1}–10^{-2}. It can be seen from Table 3.11 that the values of fecal coliforms under T2 and T3 were 10^{-1} and 10^{-2}, reaching the harmless level, while those under T1 and T4 were both 1, failing to satisfy the standard.

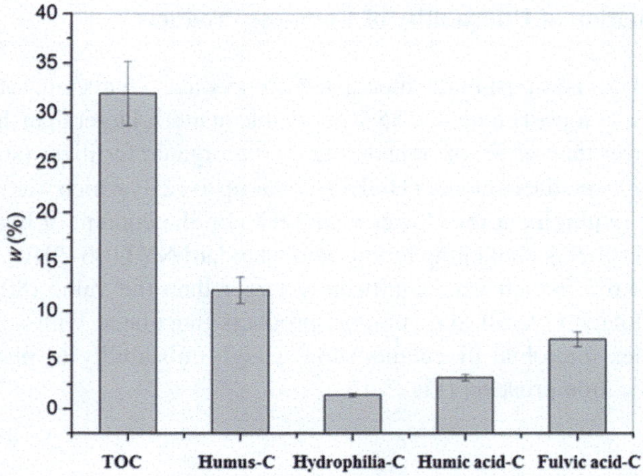

(a) The composition of organic carbon of compost

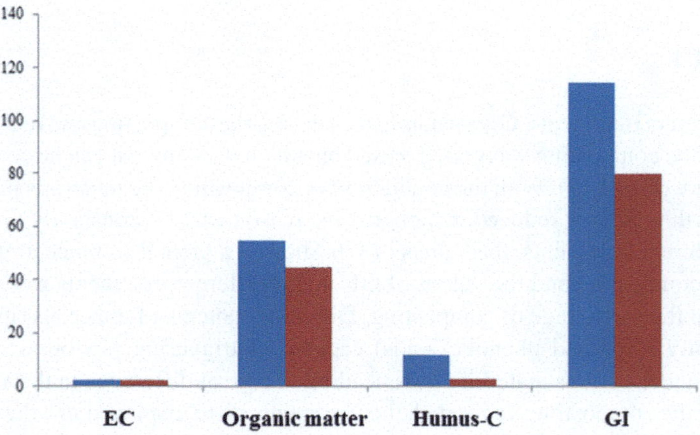

(b) Comparation of quality of compost and domestic standards for organic fertilizer

Fig. 3.16 The utilizability of compost products

3.4.2.3 Evaluation of Composting Effect

According to the evaluation index and standard of composting effect shown in Table 2.8, each treatment was subjected to weighted scoring and synthesis scores of each treatment are displayed in Table 3.12. T3 showed the highest score (7.6), followed by T2, T1 and T4. The synthesis scores of various treatments were subjected to orthogonal factor analysis. Based on the ranges of various factors, it can be seen that biogas residue intensively influenced the fermentation effect of mixed compost while chicken and pig manures showed an insignificant influence. By comparing the average effects of biogas residue, pig and chicken manures at different levels, it can

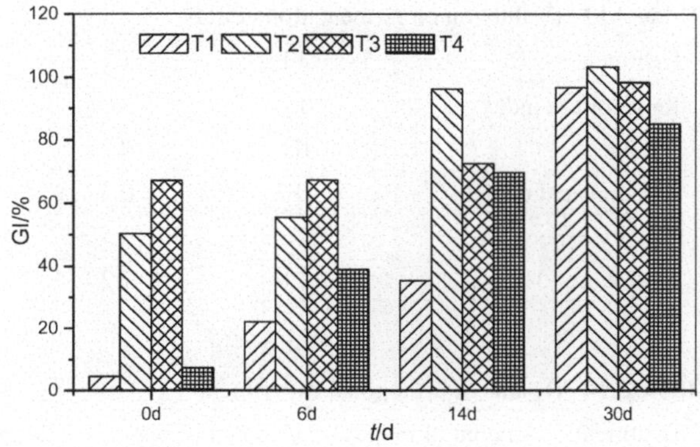

Fig. 3.17 Changes of GI of different treatments during composting

Table 3.11 Fecal coliform value after composting

Groups	Fecal coliform value
T1	1
T2	10^{-1}
T3	10^{-2}
T4	1

Table 3.12 Evaluation standards for composting effect

Evaluation index	Scoring standard
Temperature (3 points)	(1) Temperature does not rise to 50 °C and therefore fermentation fails, having score 0; (2) Heating rate is fast and high-temperature period lasts for a long time. Moreover, after stepping into the secondary fermentation, there is a high temperature, with a favorable heat preservation effect, having score 3; (3) When the condition is between the aforementioned two aspects, it scores 1 or 2 points based on the heating rate during fermentation or temperature of secondary fermentation
T value of compost samples (3 points)	T values of samples under each treatment are displayed in a descending order and the product of the rank and 0.3 is taken as the score of the treatment
Sensory effect (3 points)	(1) The compost materials are quite distinct with each other, failing to be mature, and the compost emits highly odorous gas, having score 0; (2) Straw is slightly mature and the compost has greatly unpleasant odor, having score 2; (3) The compost materials are sufficiently mixed and the compost emits odor that can be tolerated, scoring 3 points
Total nutrient content (1 point)	After the composting is ended, the total nutrient contents of samples under each treatment are shown in a ascending order and the product of the rank and 0.1 is taken as the score of the treatment

Table 3.13 Evaluation on fermentation effect

Treatment	T1	T2	T3	T4
Temperature index	1	3	3	1
T value	0.6	0.9	1.2	0.3
Total nutrient content	0.2	0.3	0.4	0.1
Sensory experience	2	3	3	1
Total score	2.8	7.2	7.6	2.4

Table 3.14 Results of orthogonal experiment

Treatment	Amount of biogas residue	Amount of pig manure	Amount of chicken manure	Synthesis score
T1	7.02	7.425	8.19	3.8
T2	5.85	7.425	6.825	7.2
T3	5.85	8.49	8.19	7.6
T4	7.02	8.49	6.825	2.4
Mean 1	3.100	5.500	5.7	
Mean 2	7.400	5.000	4.8	
Range	4.300	0.500	0.900	

be seen that the average effects were higher than 7.02, 8.49, and 6.825 when the biogas residue, pig manure, and chicken manure were at the levels of 5.85, 7.425, and 8.19, respectively. On the basis, the optimal proportion of mixed ratio–biogas residue, pig and chicken manures was 5.85:7.425:8.19 (Tables 3.13 and 3.14).

3.5 Research of Bio-augmented Composting Equipment of Organic Wastes

3.5.1 Equipment for Quick and Resourceful Treatment of Kitchen Wastes

Product modularization is an effective method to support self-designed products of users. Module is a standard part with independent function and can be input and output. Parts generally include sub-units, assembling units, components, etc. Adopting modular design, the equipment mainly aimed to better realize standardization, generalization and serialization and facilitate production and maintenance in late stage. In addition, the design reduced resource waste and increase of time and cost caused by maintenance and replacement, reflected the idea of green design and strengthened market power.

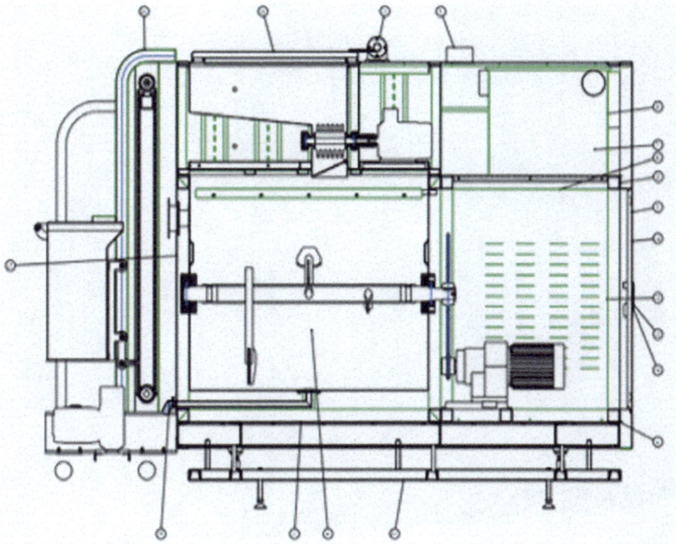

Fig. 3.18 Profile map

Based on patented technologies of "A quick and resourceful treatment equipment of kitchen wastes (ZL201410471741.8)" and "A sorting device and method for resourceful utilization products of household wastes (ZL201210490644.4)", this study researched an integrated bio-augmented composting equipment of organic wastes. As displayed in the Figs. 3.18, 3.19 and 3.20, the equipment with net dimension (excluding the base) of 1.9 m × 1.4 m × 0.9 m (length × height × width) comprised an automatic feeding system, a crushing system, a main transmission system, a master control system, a purification system and an auxiliary system. Furthermore, the equipment had the unique cutter design, which effectively avoided the wrapping of plastics. The secondary speed reduction design ensured stable transmission and automatic feeding and realized process automation. The system showed low energy consumption and could control secondary pollutions. The processor adopted the integrated, modular design concept, so each system could be independently operated and controlled and showed advantages, such as good timeliness, accurate control, simple operation, energy saving and environmental protection (Table 3.15). Therefore, it has been widely used.

3.5.2 *Harmless Treatment Equipment of Dead Livestock and Poultry*

In recent years, animal husbandry has developed rapidly in China which has been a country with the largest livestock and poultry breeding and the greatest consumption

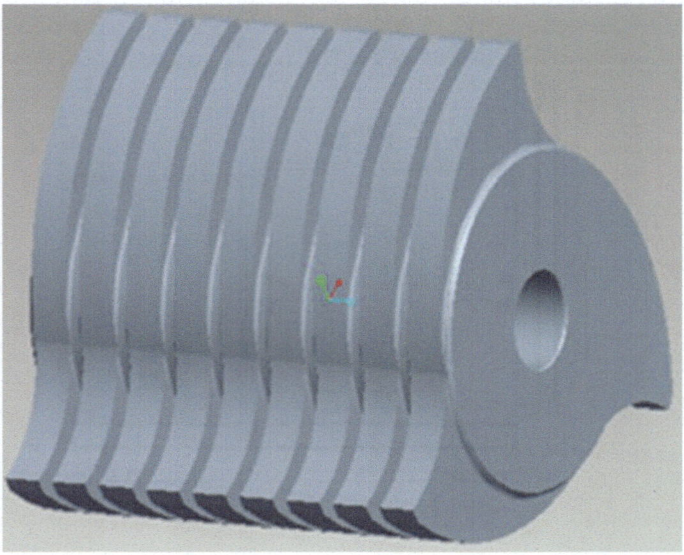

Fig. 3.19 Outside view of crushing tool

Fig. 3.20 Quick resourceful treatment equipment

Table 3.15 Main technical parameters of the integrated bio-augmented composting equipment of organic wastes

Model	HX300
Total power (kw)	7.5
Capacity (kg)	300
Weight (kg)	1000
Voltage (V)	380
Length (m)	2.48
Width (m)	1.73
Height (m)	1.0
Service life (y)	15
Noise reduction system	With
Deodorization system	With
Sensor monitoring	Temperature, humidity, oxygen content, air volume
Material	Stainless steel 304
Operation mode	Continuous/intermittent

in the world. However, some drawbacks leading to serious deterioration of ecological environment have become increasingly prominent. In particular, these problems, such as a lot of excrements and various wastes produced by breeding as well as dead livestock and poultry have gradually threatened animal husbandry production, public health and even meat safety. Therefore, how to prevent and control pollution in livestock and poultry breeding and realize harmless treatment and resourceful utilization of wastes is an important bottle neck problem urgently needing to be solved for sustainable development of livestock and poultry breeding industry in some provinces even the whole nation. The equipment for harmless treatment of dead livestock and poultry combined high-temperature gasification with biological fermentation and integrated autonomy patents, such as automatic feeding by robots, visual touch-screen operation and sterilization in high-voltage electric fields. The equipment had dominant advantages, such as high degree of automation, no smoke or noise, no environmental pollution, simple operation and low maintenance costs. By adding crop straws and chaffs during the harmless treatment as supplementary materials, the produced compound fertilizers had high quality, stable compositions and performances and no pathogenic bacteria and could be directly applied to soils, thus showing important environmental and economic benefits.

3.5.2.1 Product Structure

1. Main body: All structures of main body of the equipment adopted high-quality 304 stainless steels with the thickness of 6 mm.

2. Rack: The rack was welded with Q235A high-quality box iron and angle iron and enabled the utilization of the equipment through strict quality control and specific process requirements.
3. Cutter: The cutter combined activity with inertia, which was good for cutting and smashing. The tool rest was manufactured of 304 stainless steel, while the blade was made of wear-resistant 65Mn steel. The tool rest and the blade were connected by using bolts, which solved the problem of replacement after wear and failure. When the blade is worn, it is just necessary to adjust corresponding bolts for replacement and maintenance, instead of replacing the entire cutter, which could improve work efficiency and save costs. This cutter has obtained the authorization of a national utility model patent (patent number: ZL 2014 2 0847515.0).
4. Heating system: Based on the principle of electric-heating, the equipment was heated using high-quality heat transfer oil by transferring the oil to the surrounding of the container through a hot oil pump. The heating system has obtained the authorization of a national utility model patent with the number of ZL 2014 2 0847513.1.
5. Electrical part: All of the control parts were purchased from internationally well-known brands and the main engine used Siemens frequency conversion technology, so as to ensure long-time and high-efficiency operation with low energy consumption. Siemens programmable logic controller (PLC) was utilized to control each equipment, so that programs are controlled and operated automatically. All commands could be sent and controlled through the touch-screen operation, so the operation was simple and convenient. Moreover, there were emergency stop and failure indication, so as to judge in advance and deal with in time when meeting emergency. All low-voltage components were produced by Schneider. According to the customer demands, it had real-time monitoring function on mobile phones (Fig. 3.21).

3.5.2.2 Relevant Parameters and Indexes

See Table 3.16.

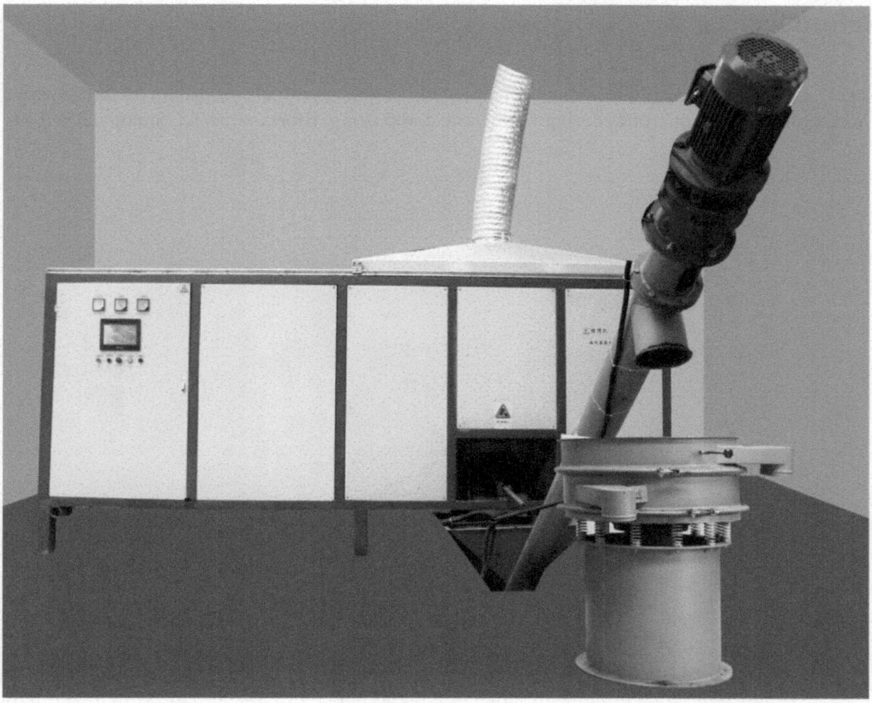

Fig. 3.21 Schematic diagram of products

Table 3.16 Equipment parameters

Item	Parameter		
	14FW-500	14FW-1000	14FW-3000
Installed capacity (kg)	750	1500	4000
Batch treating capacity (kg)	500	1000	3000
Batch treating time (h)	≤48	≤48	≤48
Batch power consumption (kW h)	120	300	820
Highest temperature (°C)	140	140	140
Power (kW)	2.2	5.5	15
Voltage (V)	Three-phase AC380	Three-phase AC380	Three-phase AC380
Overall dimension (mm)	3000 × 1200 × 1600	4000 × 1200 × 1800	6000 × 2000 × 2500

Reference

Song C, Li M, Qi H, et al. (2018). Impact of anti-acidification microbial consortium on carbohydrate metabolism of key microbes during food waste composting. Bioresource Technology, 259:1–9.

Chapter 4
Development of the Standard System for Differentiated Resourceful Utilization of Organic Wastes in Villages and Towns

Abstract The Outline of the National Medium- and Long-term Plan for Scientific and Technological Development (2006–2020) clearly regards the implementation of technical standard strategy as one of the two major strategies for scientific and technological development in China. Moreover, it is the challenge and requirement for standardization in the face of the international situations and domestic development. Actions of environmental protection, such as major scientific and technological projects for comprehensive environmental improvement in villages and towns and continuous improvement have been taken. Based on this, a series of standards and specifications, such as resourceful treatment of domestic wastes and livestock manure in villages and towns have been compiled. The Ministry of Environment and Ecology of the People's Republic of China has recognized the above work. The ministry pointed out that Ningxia Hui Autonomous Region has formulated and promulgated a series of standards for comprehensive improvement of rural environment through active innovation based on the actual conditions in the promotion of continuous improvement of rural environment. This provides guarantees for doing all tasks well and the practices and experience are worth learning and promoting. The promulgation of standards and norms provides an important basis for rural environmental protection in Ningxia and ensures that continuous improvement of rural environment is efficiently and orderly carried out. Relevant achievements have award the second prize of National Excellent Engineering Consultation.

4.1 Overview

The Outline of the National Medium- and Long-term Plan for Scientific and Technological Development (2006–2020) clearly regards the implementation of technical standard strategy as one of the two major strategies for scientific and technological development in China. Moreover, it is the challenge and requirement for standardization in the face of the international situations and domestic development. Actions of environmental protection, such as major scientific and technological projects for comprehensive environmental improvement in villages and towns and continuous improvement have been taken. Based on this, a series of standards and specifications,

M. Li et al., *Differentiated Resourceful Utilization of Rural Organic Wastes*,
https://doi.org/10.1007/978-981-15-2712-8_4

Fig. 4.1 Technical standard seminar

such as resourceful treatment of domestic wastes and livestock manure in villages and towns have been compiled. The Ministry of Environment and Ecology of the People's Republic of China has recognized the above work. The ministry pointed out that Ningxia Hui Autonomous Region has formulated and promulgated a series of standards for comprehensive improvement of rural environment through active innovation based on the actual conditions in the promotion of continuous improvement of rural environment (Fig. 4.1). This provides guarantees for doing all tasks well and the practices and experience are worth learning and promoting. The promulgation of standards and norms provides an important basis for rural environmental protection in Ningxia and ensures that continuous improvement of rural environment is efficiently and orderly carried out. Relevant achievements have award the second prize of National Excellent Engineering Consultation.

4.2 Operation Norms of Facilities for Treating Rural Domestic Wastes (DB 64/T 867-2013)

4.2.1 Application Scope

The standard specifies general regulations on the operation norms of facilities for treatming rural domestic wastes and the operation technologies of equipment for waste collection and transportation, waste landfill sites, and composting equipment.

The standard is applicable to the operation management, maintenance, and safety operation of facilities for disposing rural domestic wastes.

4.2.2 *Standards for Reference*

The following documents are essential for the application of the standard. The documents with dates only refer to those versions issued on the specific dates. For the undated documents, only the latest versions (including all of the modification lists) are applicable.

Safety color (GB 2893)
Safety signs (GB 2894)
Environmental quality standard for noise (GB 3096)
Safety regulation for railway and road transportation in plants of industrial enterprises (GB 4387)
General principals for the requirements of safety and health in production process (GB/T 12801-2008)
Standards for pollution control on the *storage and* disposal site *for* general industrial solid wastes (GB 18599)
Code for design of extinguisher distribution in buildings (GB 50140)
Hygienic standards for the design of industrial enterprises (GBZ 1)
Leachate treatment project technical specification of municipal solid waste landfill (HJ 564, tentative)
Regulations on safety work in electric power industry (DL 408)
Technical code for sanitary landfill of municipal domestic refuse (CJJ 17, including clause explanation)
Technical specification for operation and maintenance of municipal domestic refuse sanitary landfill (CJJ 93, including clause explanation)
Regulations on hazardous wastes in the Ningxia Hui Autonomous Region (issued by Government of the Ningxia Hui Autonomous Region in April, 2011).

4.2.3 *General Regulations*

4.2.3.1 Operation Management

1. Operating personnel on each post should be familiar with the relevant treatment process, responsibility and quality requirement of their own posts. Operating personnel should be acquainted with the technical performance, operation and maintenance, and safety operation rules of the equipment and facilities of their own posts. Personnel on special posts need to have vocational training.
2. Operating personnel should stick to their posts and make careful record. Management personnel need to check the operation conditions of facilities, equipment, apparatuses, and instruments regularly. Once abnormalities are found, the corresponding responses and measures should be taken and this needs to be reported to the higher authorities layer by layer.

3. Technical files specifying the utilization and maintenance of all kinds of mechanical equipment, apparatuses, and instruments are expected to be constructed at each treatment sites of rural domestic wastes. Standardized management should be applied for data about each technique and operation record.
4. The operation management, maintenance, and safety operation of electrical substations should accord with the requirements in standard DL408.

4.2.3.2 Maintenance

1. Facilities such as roads and drainage system should be checked and maintained at regular intervals. Once abnormalities are found, they should be repaired in time.
2. Power supply facilities, electric appliance, lighting equipment, and communication lines need to be checked and maintained regularly.
3. Loading and lifting equipment in waste transfer stations, landfilling equipment, facilities for mechanical separation, and transport vehicles should undergo routine and regular graded main.
4. Lightning-protection and explosion-proof equipment should be inspected and maintained by professional according to relevant standards.
5. All kinds of transportation and notice signs need to be inspected and replaced regularly.
6. All kinds of fire-fighting devices should be checked and replaced regularly.

4.2.3.3 Safety Operation

1. The safety and healthy management during the operation should obey the regulations in standard GB/T 12801-2008.
2 The setting of fire-fighting devices needs to accord with regulations in standard GB 50140.
3. Striking signs should be set in locations prone to disasters, in accordance with the regulations in standards GB 2893 and GB 2894.
4. Transportation management in the collection and disposal sites for domestic wastes should follow the regulations in standard GB 4387.
5. Regulations on hazardous wastes in the Ningxia Hui Autonomous Region should be followed concerning the collection, storage, transportation, and disposal of hazardous wastes such as inflammable, explosive, and highly toxic ones.

4.2.4 *Operation Technologies of Waste Collection and Transportation Equipment*

4.2.4.1 Operation Management

1. Waste collection equipment

The collecting boxes (or pools) should be matched with the waste transfer tools.

The waste transfer containers should be sealed to avoid secondary pollution during the transportation.

2. Waste transfer stations

The stacking and storage of wastes in the transfer stations require to accord with regulations in standard GB 18599.

The unloading and loading positions in the transport workshops should allow the return of vehicles.

The dispatch of the whole transfer station should follow the technological process and be operated orderly.

Specific personnel should be assigned to specific posts in the control room and others are not allowed to enter the room and use the equipment therein without permission.

The functions of equipment and facilities in the management room are not allowed to be changed, and tools should be put at given positions.

The transfer vehicles and presses should be matched.

It should be guaranteed that the waste is transferred on conditions with effective control over the pollution in a relatively closed environment.

Waste transfer workshops should be equipped with folding doors that are easily opened and closed and have normally closed vents.

According to the technical design of the waste transfer unit, the ventilation, dust prevention, and deodorization measures should be strengthened at key positions such as the loading and unloading locations of wastes in the transfer stations. Large transfer stations need to be equipped with independent air ventilation and deodorization systems.

Place around waste transfer stations needs to be clean and has no waste and sewage, and presents favorable landscaping. Routine fly eradication and disinfection should be carried out, ensuring that there are no eyesores in front of and within the sight of a transfer station.

Ground should be plat in waste transfer stations with no deposited sludge retained and arranged with sewage channels (or pipes). Meanwhile, effective sewage treatment measures should be taken to avoid secondary pollution.

3. Waste transportation

While rickshaws are used for collecting wastes, the radius for collection should be within 1 km, not larger than 2 km.

While mini-sized motor vehicles are used for collecting wastes, the radius for collection should be lower than 3 km, not larger than 5 km.

When mini-sized flat trucks are utilized for transferring the collected wastes, the flat trucks have to enter the transfer stations to unload wastes, so as to avoid secondary pollution.

Wastes are loaded immediately after a flat truck arrives to shorten the retention and exposure time of wastes.

Waste transportation vehicles should have favorable sealability on the whole. No waste scattering and sewage dripping are allowed during the transportation. The vehicles should be clean and on good conditions.

Drivers of waster transfer vehicles should check whether the inlet and outlet in front and back of the sewage box are blocked frequently in rainy season.

Dustbins should be cleaned while they are lifted to transfer wastes to the vehicles and the transfer stations should be weeped after departure of transfer containers.

4.2.4.2 Maintenance

1. Dustbins should be repaired in time to guarantee normal utilization.
2. After transferring wastes in dustbins to waste transfer vehicles, drivers should get off to check whether all the wastes are unloaded and clean frame of the unloading outlet of the dustbins. No wastes are allowed to be retained on the frame.
3. Waste transfer vehicles should be clean before leaving the transfer stations.
4. Locations including the press, trough, management room, tool room, operation platform, and so on should be kept clean.
5. The air cleaning and sewage treatment facilities are expected to be checked and cleaned regularly.
6. Waste transfer stations should be kept clean with clear signs.

4.2.4.3 Safety Operation

1. The safety and occupational health of personnel working in the waste transfers stations should accord with regulations in standards GB/T 12801-2008 and GBZ 1.
2. The noise control during the waste transfer process should conform to regulations in standard GB 3096.
3. Safety signs such as instructions of traffic control and open-fire control should be set at appropriate positions in waste transfer stations.
4. Warning signs need to be set on relevant facilities and equipment for waste loading, lifting, and reversing vehicles.
5. Directing and positioning devices or limitation pre-warning devices should be arranged at the loading and unloading positions according to the specifications of waste transfer vehicles or containers.

6. Safety of cleaning personnel should be guaranteed during the cleaning and safety warning marks need to be set on the waste transport vehicles.

4.2.5 Operation Technologies of Waste Landfill Sites

4.2.5.1 Operation Management

1. Landfilling operation

Plans and schemes of landfilling operation should be formulated prior to waste landfilling. Wastes should be filled layer by layer in each unit and region.

The waste landfilling platform should get prepared everyday before the landfilling operation and the area of the platform needs to be controlled according to the actual conditions.

Specially-assigned personnel should be responsible for the command and dispatch of vehicles at the landfilling sites.

Wastes need to be paved in time during the landfilling and each layer of paved wastes should be lower than 1 m. It is suggested to set the thickness of a unit to be 2–3 m.

Hazardous wastes are not allowed to be filled in waste landfill sites. Bulky wastes that can probably form cavity structures should be crushed in advance.

The waste landfill sites should be covered in time after daily operation. The covering layer should be compacted and smoothed. The thickness of daily and intermediate covering layers should be not less than 15 and 20 cm, respectively. The thickness of the final covering layer needs to be set in line with the requirements for site sealing.

Drainage ditches, water cut-off and diversion dams, and downstream ditches in waste landfill sites should be kept unblocked.

2. Operation of ventilation system

During the collection and growing of gas in waste landfill sites, deposited sludge and sundries should be cleaned in time to guarantee that pipes in the well are connected smoothly and devices in air ventilation system are under good conditions.

3. Collection of leachate

Collection system for leachate in waste landfill regions should be kept unblocked.

Leachate in the waste landfill regions needs to be monitored regularly.

Leachate should be harmlessly treated, and not allowed to be discharged directly. The treatment and discharge of leachate need to accord with regulations in the standard HJ 564.

Leachate needs to be sprayed timely for dust prevention.

4. Mechanical operation of landfilling

Routine inspection and maintenance should be conducted for the operation machines before landfilling operation.

Before operating landfilling machines, all instruments should be checked to ensure working normally. Once abnormalities are found, the machines should be halted immediately for inspection.

Landfilling machines should be managed with assignment of specified vehicles and personnel and the shift relief system should be applied.

Wastes and sundries on the landfilling machines should be removed timely after finishing landfilling operation.

The inspection and supervision of wind-break nets should be strengthened and wastes hung on the nets need to be removed in time.

5. Environmental monitoring and operation supervision of waste landfill sites

Environmental monitoring and evaluation should be carried out regularly after the operation and closure of waste landfill sites.

Upstream and downstream water quality as well as environmental and air quality around the waste landfill sites should be monitored once every half a year.

From the beginning of landfilling operation to the ending of site closure, subsidence of the wastes should be monitored once annually.

6. Disinfection

Rodenticide, culicide, and muscicide of low toxicity, high efficient, and high pertinence should be used regularly in the waste landfill sites and other areas where there are abundant mosquitoes, flies, and mouses for eradication.

7. Site closure

The closure of waste landfill sites should conform to the regulations in standards CJJ 17 and CJJ 93.

After closure of waste landfill sites, facilities for drainage, air ventilation, transportation, and leachate treatment in the sites need to be operated and managed according to the design requirement.

4.2.5.2 Maintenance

1. Specified personnel should be assigned in the waste landfill sites responsible for the maintenance, clean keeping, and desilting of facilities including roads, cut-off ditches, drainage channels, flood barriers, waste dams, and car-washing tanks in the landfill region.

2. The protecting layers for side slops in the sites and facilities for seepage-proofing, drainage, and so on in the regions to be filled should be regularly checked and maintained.
3. Pipes, wells, and pools in the waste landfill sites need to be checked and maintained regularly.
4. If machines need to be retained in waste landfill sites for more than one week, corrodible parts such as tracks and compacting teeth should be subjected to anticorrosive and rust-proof treatment.
5. Eradication devices and drugs should be maintained and supplemented regularly, respectively.
6. Regular check and maintenance needs to be conducted for facilities such as monitoring wells in the waste landfill sites.
7. Facilities in the waste landfill sites should be clean, and the ground is required to be clean and flat.

4.2.5.3 Safety Operation

1. Safety warning marks should be set in the waste landfill sites.
2. Operation and management personnel need to have safety education and manoeuvre for fire prevention and explosion prevention regularly and be regularly inspected and examined.
3. No enclosed buildings and structures are allowed to be constructed in the operation region of waste landfill sites.
4. Safety measures should be taken as soon as possible once the methane concentration exceeds the limit in the sites.
5. Machines should be operated without faults.
6. Specific personnel should be on duty during heavy rain and storm to inspect the drainage of the drainage system. Once the facilities are found damaged or blocked, personnel should be organized in time to dispose the issue.

4.2.6 Operation Technology of Composting Equipment

4.2.6.1 Operation Management

1. Mechanical separation

Before operation of plate-type feeders, the material inlet should be checked in case it is clamped. The operating conditions of the material inlet and the electromechanical device should be monitored continuously during the operation process. In case of a fault, the machine should be turned off immediately and maintained.

Belt conveyors should have borders higher than 15 cm.

Belt conveyors should be adjusted in time during the operation when the following phenomena happen, such as belt deviation and scattering of wastes to guarantee the continuous and smooth operation.

Before the operation, rotary screens should be checked to ensure that there is no material in the screen drum, the screen surface is unblocked, cutters are not twined by sundries, motors and transmission devices are on good conditions, and rollers are undamaged, deviated, or loosed.

During the operation, it needs to check and ensure that the materials are loaded smoothly and the motors or bearings do not have excessive temperature rise.

Residual materials in the screen drum should be removed timely after the accomplishment of the screening.

Once abnormalities are found, the machine should be turned off timely for maintenance. After clearing the fault, the machine should idle for 3–5 min before being operated.

2. Composting sites

Facilities for dust prevention, deodorization, fly eradication, disinfection, and sewage collection should be arranged in the composting sites. The ground in production regions should be hardened. The allowable concentrations of odor gases, such as hydrogen sulfide, sulfur dioxide, and ammonia need to accord with the regulations in standard GBZ 1.

Phenomena including non-uniform material thickness and moisture content, and material compaction that unfavorable for the fermentation and temperature rising should be avoided.

The height of waste dumps for static-pile composting with natural ventilation should be in the range of 1.2–1.5 m. The height can be increased to 2–3 m in the case that ventilating ducts are set at the bottom of the dumps.

For forced-aeration static-pile composting, the air flow should be 0.05–0.20 Nm^3 per minute per cubic meter of wastes. The air volume should be adjusted in accordance with the requirement of composting process.

The temperature change in the composting sites should be detected everyday. The temperature monitoring points need to be set on different layers and regions according to the temperature rising and change law.

The equipment and staff passages in each operation region of the composting sites should be kept unimpeded.

3. Supporting facilities

Main operation parameters of wind blowers during operation such as air volume and air pressure need to be recorded.

Rainwater and sewage in the drainage system of the composting sites should be shunted. The reflux ditches need to be cleaned and dredged timely to ensure leachate to smoothly flow to the sewage pool.

4. Control and monitoring

The control and monitoring instruments of process facilities should be checked to ensure they are in good condition before running the facilities.

The control room (or monitoring site) should have a good sight so as to observe and control the operations of relevant processes and equipment.

The control room should inform relevant information of an accident to the processes before and after the process attacked by the accident.

4.2.6.2 Maintenance

1. Residues on transmission parts (friction wheels or gears) of rotary screens should be removed in time. The screen surface needs to be cleaned, repaired, and replaced regularly.
2. Wind blowers and motors should be maintained regularly; filter bowls, filter screens, and filter bags need to be regularly cleaned, maintained, or replaced; lubricating oil (or grease) should be smeared on rotating parts such as bearings at regular intervals.
3. Gutterways, ventilating ducts, and gutter cover plates on the bottom of composting sites should be cleaned and dredged regularly.

4.2.6.3 Safety Operation

1. Operation record needs to be checked before running equipment including plate-type feeders and rollers, to ensure that the motors and speed-regulation devices are on normal conditions and the whole machines and their transmission parts are not clamped.
2. Plate-type feeders and rollers need to be turned off for inspection and maintenance during operation if the following phenomena occur: abnormal noise, faults of parts such as breakage, too high temperature of motors or bearings, and clamping of material inlets and outlets.
3. Once machines are clamped, blocked, or twined by sundries, they need to be turned off immediately to eliminate the faults.
4. Operating personnel are not allowed to pull foreign matters clamping, blocking and twining machines before turning off the machines.

4.3 Technical Code for Livestock Manure Composting (DB 64/T 871-2013)

4.3.1 Application Scope

The standard specifies the terms involved in livestock manure composting technology and their definitions, and requirements of site selection, technological process, pre-treatment, control over composting technology, ancillary projects, product packaging, signs, transportation, storage, environment, safety management, and inspection.

The standard is applicable to livestock manure composting facilities and technological process with daily capacity larger than 1 t.

4.3.2 Standards for Reference

The following docments are essential for the application of the standard. The documents with dates only refer to the versions issued *on the specific dates*. For the undated documents, only the latest versions (including all of the modification lists) are applicable.

Safety regulation for railway and road transportation in plants of industrial enterprises (GB 4387)
Integrated wastewater discharge *standard*
Emission standard for industrial enterprises noise at boundary (GB 12348)
Emission standard for odor pollutants (GB 14554)
Code of design on building fire protection and prevention (GB 50016)
Code for design of extinguisher distribution in buildings (GB 50140, including clause explanation)
Hygienic standards industrial enterprises (GBZ 1)
General principles for the requirements of safety and health in production process (GB/T 12801-2008)
Organic fertilizer (NY 525)
Technical code for overall plane design of livestock farms (NY/T 682-2003)
Compound microbial fertilizer (NY/T 798-2004)
Microbial organic fertilizers (NY 884)
Determination of atmospheric articles PM10 and PM2.5 in ambient air by gravimetric method (HJ 618).

4.3.3 Terms and Definitions

The following terms and definitions are applicable to the current standard.

4.3.3.1 Pre-treatment

It is a treatment process to modify the fermentation conditions of raw materials of livestock manure compost by means of mechanical crushing, addition of ingredients, and inoculation of mixed animalcule bacteria.

4.3.3.2 Primary Fermentation

It is the degradation process of putrescible organic matters in the raw materials of compost experiencing temperature rising, high-temperature treatment, temperature drop, and temperature stabilization.

4.3.3.3 Secondary Fermentation

It refers to the further maturation process of compost after the primary fermentation, also known as the aging or composting stage.

4.3.3.4 Aftertreatment

It refers to the fine treatment of maturated compost, involving screening, crushing, and drying processes.

4.3.4 Project Construction and Site Selection

4.3.4.1 Construction of Composting Plants

The primary projects include the construction of pre-treatment, primary fermentation, secondary fermentation, aftertreatment, and measuring and packing regions. The ancillary projects involve loadometer, sewage treatment, product inspection, finished product warehouse, and fire protection.

4.3.4.2 Site Selection

1. The site selection should accord with requirements of development planning of village and town construction, development planning of land utilization, and the environmental protection planning.
2. The site should be reasonably distributed by taking into account the location characteristics of livestock and poultry farms (villages) and taking full advantages of compost treatment facilities that have been built and plan to be built.
3. The safety protection distance for treatment facilities of manure and compost in livestock and poultry farms (villages) should follow the regulations in standard NY/T 682-2003.
4. The livestock manure composting plants and treatment facilities are not allowed to be built in the following regions:

    (a) Conservation areas for sources of living and drinking water, scenic sites, natural reserves, and residential areas;
    (b) Areas threatened by flood or torrential flood and frequently attacked by natural disasters such as debris flows and landslides;
    (c) When constructing treatment facilities for livestock manure near to the areas not allowed building composting plants, the facilities should be built at the downwind or lateral downwind direction of the prevailing wind direction all the year round. The plant boundary should distance at least 2000 m to the boundary of the areas; and
    (d) Treatment facilities of livestock manure should be 500 m farther from surface waters.

4.3.5 Technological Process

Current composting processes include static-pile composting, forced-aeration static-pile composting, composting with mechanical turning and agitation, and combined processes. The technological process is illustrated in Fig. 4.2.

4.3.6 Pre-treatment

4.3.6.1 Before composting of livestock manure, agricultural residues such as straw, bran coat, soybean meal, and mushroom bran are suggested to be added as ingredients to adjust the moisture content and carbon–nitrogen ratio (C/N) of the raw material of the compost. In addition, the materials need to be crushed to ensure that the raw materials meet the requirement of fermentation. The pre-treated raw materials of compost should meet requirements in Table 4.1.

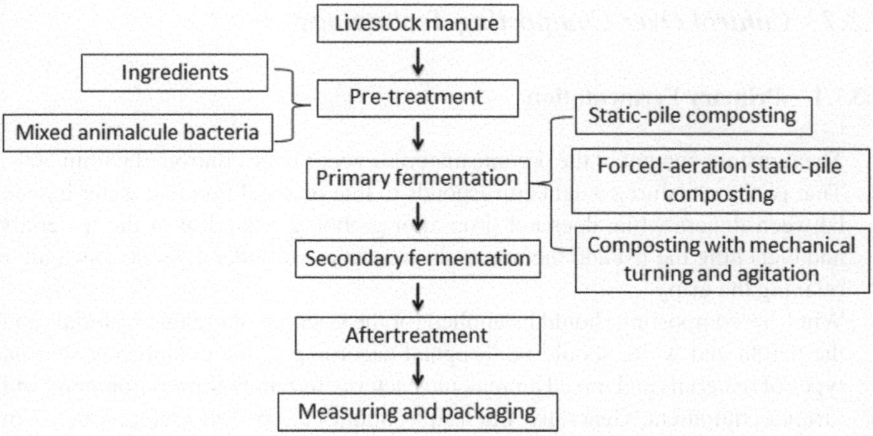

Fig. 4.2 Flow chart of livestock manure composting technology

4.3.6.2 Mixed animalcule bacteria should be mixed and stirred evenly with the original ingredients according to a certain proportion so as to inoculate them in the raw materials. The inoculum size should conform to the utilization requirement of the bacteria. Relevant regulations in standard NY/T 798-2004 should be followed for the inoculation of the mixed animalcule bacteria.

4.3.6.3 The materials should be uniformly and loosely distributed, avoiding the occurrence of non-uniform thickness and moisture content.

4.3.6.4 Following materials are prohibited to be added in raw materials of compost:

(a) Poisonous and harmful industrial products and their wastes, and municipal sludge;
(b) Materials likely to have chemical reactions and produce poisonous substances;
(c) Corrosive or radioactive materials;
(d) Hazardous substances such as inflammable and explosive materials;
(e) Biohazardous substances and medical wastes;
(f) Microbial agents harming the environment; and
(g) Other non-degradable solid wastes.

Table 4.1 Control indexes of raw materials of the compost

Term	Index
Moisture content	50–65%
Degradable organic matter (dry basis)	≥45%
Particle size	≤60 mm
Carbon–nitrogen ratio (C/N)	25:1–35:1

4.3.7 Control Over Composting Technology

4.3.7.1 Primary Fermentation

1. The moisture content of the organic materials needs to be controlled within 60%. That is, the moisture content corresponds to that on condition that water is seen between fingers while does not drop after grabbing a handful of the materials and clenching the fist and the materials are dispersed under a gentle touch after relaxing the grip.
2. Windrow composting should be applied for the stacking of organic materials and the height and width should be designed according to the composting season, types of materials and mixed animalcule bacteria, fermentation environment, and turning equipment. Generally, the heaps should be 0.6–2 m high and 0.8–2 m wide.
3. During the fermentation, temperature of heaps should be measured three or four times every day. The measuring points should be inside the heaps 10–30 cm to the surface. The compost should be kept at 55 °C for 5–7 days to reach the harmlessness standard and the highest temperature should not exceed 70–75 °C (applying the temperature at which mixed animalcule bacteria are killed as the limit).
4. After exceeding 60 °C and keeping the compost at the temperature for 48 h, the heaps should be turned. The heaps need to be turned once every three to five days and turned immediately after the temperature exceeds 70–75 °C. The heaps should be turned sufficiently and evenly, to bring the bottom materials into the upper and middle part of the heaps, so that they can be sufficiently decomposed.
5. In the forced-aeration static-pile composting, the air flow should be 0.05–0.20 Nm^3 per minute per cubic meter of materials. The air pressure should be increased by 1–1.5 kPa every rising the height of the material layer by one meter.
6. The period of primary fermentation generally should be longer than 15 days.
7. As the fermentation terminates, the temperature of materials does not rise any longer and the heaps are basically not smelly and approximately grey brown.

4.3.7.2 Secondary Fermentation

1. Fresh raw materials of compost are not allowed to be added in the secondary fermentation.
2. The moisture content should be controlled within 40–50%.
3. To reduce nutrient loss, the temperature of materials is suggested to be controlled below 50 °C. The temperature of heaps can be controlled by adjusting the height of material layers.
4. The value of pH needs to be within 5.5–8.5. If it is beyond this range, adjustment measures should be taken.
5. The period of secondary fermentation basically lasts for 15–30 days.

6. After termination of the fermentation, the maturated compost should satisfy the following requirements:

   (a) The compost is brown or grey brown, loose, and scentless, and does not contain mechanical admixtures;
   (b) The moisture content is blow 30%;
   (c) The carbon–nitrogen ratio (C/N) is smaller than 20:1; and
   (d) The oxygen consumption rate is basically stabilized.

4.3.7.3 Aftertreatment

1. The fully decomposed and stable compost products should be crushed, screened, dried, and granulated.
2. Compost products, as a kind of organic fertilizers, should follow regulations in standard NY 525; As a kind of microbial organic fertilizers, it needs to accord with regulations in NY 884.

4.3.8 Ancillary Projects

4.3.8.1 The ancillary projects in composting plants should be matched with the primary ones.

4.3.8.2 Rainwater and sewage should be shunted using the drainage system. Independent leachate collection devices should be set up in the composting plants. After being collected, the leachate can supplement for the primary fermentation of the raw materials, or discharged after being treated with sewage treatment facilities. It is prohibited to discharge the leachate directly.

4.3.8.3 Deodorization facilities and medicament should be used or mixed animalcule bacteria of favorable deodorization effect should be inoculated to purify and eliminate the odor gases such as hydrogen sulfide, sulfur dioxide, and ammonia.

4.3.8.4 The setting of fire-fighting equipment should meet the fire-fighting requirements and accord with relevant regulations in standards GB 50016 and GB 50140.

4.3.8.5 Inspection facilities and warehouses for finished compost products should be equipped. The storage should conform to the regulations in standards NY 525 and NY/T 798-2004.

4.3.8.6 The storage of raw materials of the compost should meet the following requirements:

   (a) Dry and wet materials need to be stored respectively; and
   (b) The ground needs to be hardened.

4.3.8.7 Environment of operating region

1. Operating region needs to be well ventilated and dustproof and dedusting measures should be taken. The emission of dust and harmful gases (hydrogen sulfide, sulfur dioxide, ammonia, and so on) should accord with the regulations in standard GBZ 1.
2. Ground of the operation region needs to be hardened and clean. No materials irrelevant to the production are allowed to be placed in the operating region.

4.3.8.8 Internal and external environments of the composting plants

1. The noise emission at boundary of plants needs to conform to regulations in the standard GB 12348.
2. The emission of leachate and sewage should follow regulations in the standard GB 8978.
3. The allowable concentration of odor pollutants at boundary of plants should conform to regulations in standard GB 14554.
4. Necessary fly eradication measures should be taken in composting plants.

4.3.9 Safety Management

4.3.9.1 The safety and sanitary management during the production should follow regulations in standard GB/T 12801-2008.

4.3.9.2 Complete regulatory framework and operating instruction for safety production need to be formulated. Operating personnel on each post should strictly execute the safety operation rules of the posts.

4.3.9.3 Safety and healthy conditions and protection appliance should be provided for operating personnel. Operating personnel should use these appliance in accordance with the regulations.

4.3.9.4 Protective covers should be used for kinematic chains of driving, transmission, feeding, and roller parts that expose to the outside of the machines, or contact prevention devices should be installed at locations exposed to operators.

4.3.9.5 Except for the parts having to be exposed, cutters in machines should be shielded or protective covers, contact prevention devices, or ejection prevention devices should be utilized.

4.3.9.6 Braking devices or emergency shutdown devices should be installed if the rotation of shaft due to inertia after poweroff is likely to lead to damages.

4.3.9.7 The transportation management in plants or workshops should follow regulations in standard GB 4387.

4.3.9.8 In order to avoid the occurrence of fire, explosion, and other serious accidents, signs of no-smoking, fire prevention, and speed limitation should be set at conspicuous positions.

4.3.9.9 The setting of fire-fighting devices should conform to regulations in standard GB 50140, inspected regularly to check the utility, and replaced in time.

4.3.9.10 Emergency response plans should be formulated for major accidents such as fire and explosion.

4.3.9.11 Safety inspection should be conducted to the whole plant at regular intervals and inspection results should be recorded.

4.3.10 Inspection Requirements

4.3.10.1 Basic requirements of inspection

1. Each batch of raw materials of compost should be inspected at least once and the terms to be inspected include moisture content, total organic matter, pH value, and so on.
2. The quality and sampling inspection of compost products should refer to relevant methods in standard NY 525.

4.3.10.2 Environment monitoring

The environment monitoring should be carried out once in the summer of each year, involving total suspended particles and odor. The inspections of total suspended particles and odor should conform to regulations in standards HJ 618 and GB 14554, respectively.

Chapter 5
Engineering Application Cases

Abstract Resource processing technology and finished products of organic waste such as kitchen waste, agricultural waste, livestock and poultry manure are widely used in many parts of China. For example, humic acid products obtained by aerobic composting of kitchen wastes are applied to improving soils, a batch of high-quality agricultural products outside is developed. In Henan, Shandong and other provinces, biogas and liquid fertilizer are produced from livestock and poultry waste. They met the standard for safety utilization of organic fertilizers and can realize high value utilization of biogas slag and slurry.

5.1 Application Case of High-Efficient Oriented Humification Technology of Kitchen Wastes

Gaoantun treatment plant for kitchen wastes in Beijing (Fig. 5.1), located in circular economy industrial park in Chaoyang district, Beijing, is one of the earliest resourceful treatment plant for kitchen wastes in China. Moreover, it was a treatment project of kitchen wastes with the largest operation scale, the highest degree of automation and the most advanced biochemical process in China. The project integrates eight process systems, such as pre-treatment, biochemical treatment, deep treatment of products, and central instrument control. The whole set of systems shows advanced technologies and strict environmental protection measures and the equipment is highly automatic. The daily treatment capacity is 400 tons and 1.32×10^5 tons of kitchen wastes are consumed annually. Moreover, 8×10^4 tons of biological humus acids are produced in each year and annual direct emission reduction of carbon dioxide reaches 1.5×10^5 tons. Agricultural carbon dioxide emissions are reduced by 9.1×10^5 tons due to reduction of coal consumption and chemical fertilizers.

Humic acid products obtained by aerobic composting of kitchen wastes are applied to improving soils, so that a batch of high-quality agricultural products bases is developed. For example, in view of problem of successive cropping obstacles of strawberry production with protected culture in Changping, Beijing, humic acid products produced by Gaoantun treatment plant of kitchen wastes in Beijing had

M. Li et al., *Differentiated Resourceful Utilization of Rural Organic Wastes*,
https://doi.org/10.1007/978-981-15-2712-8_5

Fig. 5.1 Gaoantun treatment plant of kitchen wastes in Beijing

been continuously applied to soils for planting strawberry for 10 years and basically no chemical fertilizers were used. In this way, soil organic matter increased from 1.6% to more than 3.0%. Therefore, the strawberry plantation was awarded to host the International Strawberry Symposium in 2012. In addition, humic acid products were used to improve salinized soils of paddy fields in the western Beijing, which increased soil water storage and realized disease prevention and yield increase of paddy fields. Furthermore, these products were also used to the project of improving quality of 6×10^4 mu of farmlands in Huailai county along Guanting reservoir, Hebei province, China.

5.2 Application Case of "5 + 1" Services for Improving the Quality of Farmlands in Pujiang

The project of "5 + 1" services for improving the quality of farmlands by recycling organic matter in Pujiang county, Sichuan province, China takes Pujiang county as a demonstration to build the first county-scale comprehensive agricultural service mode and a "5 + 1" service platform. The five service items include recycling of organic matter for cultivation, big data of soils, fertilizing soils, biological control and prevention and agricultural mechanization service. One back-end security platform consists of e-commerce platform, cold-chain logistics, green credit, agricultural insurance, organic certification and inspection and detection.

The project is supported by government policies. The county government has launched the planning for the development of enterprises for treating organic matter in the whole county and the three-year action plan for improving the quality of farmlands. The soil organic matter increases by 0.2 point annually. Furthermore, the government gives subsidies and guidance funds. Local governments advocate localizing third-party services and strengthening management in utilization of organic matter in wastes, source management of organic matter, and guidance of improving quality of farmlands, monitoring and evaluation.

The project is based on the three major industries in Pujiang and cultivation of healthy soils extends to 5.5×10^5 mu. The project realizes the trinity of healthy soil, quality and ecological conservation and forms a mature system that can be standardized and promoted in the whole nation.

This project consumes 6.6×10^4 tons of kitchen wastes in central urban areas in Chengdu, Sichuan province, China for each year, which basically solves disposal problem of kitchen wastes in the city with population of 2×10^6. About 7×10^4 tons of CO_2 emissions are reduced and 4×10^4 tons of biological humus acids are produced annually. Moreover, 3×10^5 tons of agricultural CO_2 emissions are reduced due to the reduction of coal consumption and fertilizers in each year. In the production, the mode combining biochemical treatment with removal of deodorization through combustion is used to control emissions of foul gases, so that harmless treatment rate reaches 100% and resource conversion rate is up to more than 95%.

Resourceful products of the project firstly pioneers public private partnership (PPP) mode for improving the quality of farmlands in Pujiang, Sichuan, a national organic agriculture demonstration county. Moreover, this project creates the "Pujiang mode" of three wins, i.e., protecting the quality of farmlands, developing agriculture and increasing incomes of peasants (Fig. 5.2). The project also takes the lead to launch the first industry fund for improving the quality of farmlands in China. At present, "Pujiang mode" has been recognised and promoted by the Ministry of Agriculture of the People's Republic of China and International Sustainable Development Foundation and is used in more than a dozen of counties in China. Moreover, the mode was awarded the fourth annual Paulson Prize for Sustainable Cities and the "5 + 1" service for improving the quality of farmlands in Pujiang was awarded the prize for project of most investment value.

5.3 Application Case of Rapid Biological Treatment of Dispersed Organic Wastes

A set of small rapid biological processing system for dispersed organic wastes has been developed and the comprehensive energy consumption of the equipment reduces by 10%. By taking agricultural wastes with high content of lignocellulose and cow dung as treatment objects, a demonstration project with the scale of 20 m^3 and volumetric biogas production of 1.56–1.75 m^3/m^3.days has been built in Shouguang

Fig. 5.2 Demonstration effects of quality improvement of farmlands

Vegetable Garden, Shandong province, China (Fig. 5.3). In addition, this technology has been promoted and applied in the projects with the scales of 50 and 100 m^3 in Xinmi, Henan province and Dianchi, Yunnan province, China, respectively.

5.4 Application Case of Energy-Saving Dry Anaerobic Fermentation Technology

The equipment of dry anaerobic fermentation technology is used for reduction, harmless and resourceful utilization of solid (high concentration) organic wastes. The equipment can produce biogases by utilizing organic matters, such as kitchen wastes, household wastes, livestock manure, treatment wastes of agricultural products and straws. The digested fermentation residues have low water content and high content of organic matter and can be directly applied to the fields after simple treatment or made organic fertilizers by further deep processing, without producing secondary pollution.

In view of the problem that the traditional wet fermentation can produce a lot of biogas slurries, the equipment can utilize vertical dry fermentation reactor with

Fig. 5.3 Scene of pilot plant test system of the complete set of vertical dry anaerobic fermentation equipment in Shouguang Vegetable Garden

high solid content in anaerobic fermentation and the feed system adopts the vertical and spiral feeding mode, which can lift materials vertically by 5 m. Conical-bottom discharge system with assembled cylindrical skirt support is used for discharge and a horizontal slurry pump is designed for reflux and distribution of biogas slurry.

This project in Zhanjiang, Guangdong province, China takes livestock manure and fruit and vegetable planting wastes as main fermentation materials and used energy-saving dry anaerobic fermentation equipment for dry anaerobic fermentation treatment in the scale of 8 m^3. Biogas produced through fermentation is converted into electricity and heat through biogas lamps and biogas stoves and supplied to the garden (Fig. 5.4). The contents of organic matters and trace elements required in crop growth in biogas fertilizers are high, which can effectively improve soil conditions, reduce plant diseases and insect pests and improve fruit and vegetable quality. Biogas fertilizers produced in the project provide nutrients for the growth of fruits and vegetables in the garden.

Fig. 5.4 Demonstration of energy-saving dry anaerobic fermentation project

5.5 Application Case of Harmless Treatment Technology of Livestock Manure

The equipment for harmless treatment of livestock manure is a high-efficiency biological bed-type processor for urine (Fig. 5.5). The equipment integrates bioengineering technologies, like rapid biological fermentation of enzyme bacteria, optimization of high-efficient microbial fermentation environment, the deodorization by microorganisms and recovery and utilization of solar and thermal energy. By using straws, weeds and rice hulls generally seen in villages as bedding materials, biological heat produced through aerobic fermentation is used for evaporating urine moisture and retaining effective nutrients. Based on this, the bad smell of wastes is removed and these including worm eggs, infectious microbe, virus and weed seeds are inactivated in 24–48 h. The treated wastes belong to high-quality farmyard manure and are rich in beneficial bacteria and bio-enzyme, which realize comprehensive treatment and resourceful utilization of planting and breeding wastes in villages. The equipment

Fig. 5.5 Equipment for harmless treatment of livestock manure

with characteristics, like rapid treatment speed, low energy consumption, small area of land occupation, no pollution and full-automation realizes unattended one-click operation.

5.6 Application Case of Comprehensive Utilization of Producing Biogas via Anaerobic Fermentation and Making Fertilizers Through Aerobic Fermentation of Agricultural Wastes

This demonstration projects is located in Shangbo village, Yifengdian county, Jimo city, Shandong province, China. This village possesses rich agricultural biomass resources and has 400 peasant households and large-scale farmlands and breeding plants nearby. Raw materials, such as cow dung and silage straws are sufficient and biogas slag and slurry can be digested. There are two CSTR with the capacity of 800 m^3 for anaerobic fermentation in the station. After separating solid and liquid fermentation residues, biogas slurry flows into the collecting pool and is used as liquid fertilizer, while biogas slag mixed with cow dung and straw in a mingled storehouse is pumped into the integrated fertilizer-producing system through aerobic fermentation with stirred biogas residues through a plunger pump (Fig. 5.6). After high-temperature fermentation for 20 days, in comparison with traditional composting, composting period shortened 15% and the content of organic matter reached 32%. Moreover, the GI was up to 90% and electrical conductivity was 2.15 ms/cm. This met the standard for safety utilization of organic fertilizers and can realize high-value utilization of biogas slag and slurry. This system runs smoothly and it feeds and discharges materials continuously and stably. With easily adjusted parameters, like temperature, water content and pH value, the system can realize full-automatic control.

Fig. 5.6 Scene of the fertilizer-producing system through aerobic fermentation with stirred biogas residues

5.7 Application of the Project of Centralised Supply of Biological Gas-I

This project in municipal demonstration area of high-tech agricultural products has built thousands of vegetable greenhouses (Fig. 5.7). In order to treat rural environmental pollution and change energy utilization structure of villages, these pollutants, such as agricultural wastes, vegetable processing wastes, excrements of human and animals and domestic organic wastes are transformed into biomass energy through anaerobic treatment. The project adopts the medium-temperature treatment process by using pretreatment and two-stage anaerobic reaction as a core and the standing time is 25 days.

After fermentation, daily biogas productions of biogas slurry and residues are 25 and 5 tons. Some biogas slurries produced in anaerobic reactor are refluxed, while others flow into the storage pool of biogas slurries and then are transported to each vegetable greenhouse used for organic vegetables. Biogas slurries after concentration are separated by a solid–liquid separator for producing organic fertilizer.

Fig. 5.7 Site and layout of biogas project

5.8 Application of the Project of Centralised Supply of Biological Gas-II

This case is a national key leading enterprise of agricultural industrialization and a provincial high-tech enterprise and possesses the largest scale of pig farm in local areas. It is also known as a test (demonstration) pig farm of the National Engineering Research Center for Livestock. By using the dry cleaning method of excrements, the pig farm in this scale produces 50 tons of solid excrements and about 270 m^3/day of washing waste water each day. Most of these excrements are directly or indirectly discharged into underdrains, which seriously pollutes surrounding environment. By building the project of biogas energy environment, the excrements are harmlessly treated to transform into energy (biogas is burned for heating or power generation) and high-quality organic fertilizer. This fundamentally solves the problem of energy and fertilizer utilization in villages and obtains good ecological, environmental and social comprehensive benefits.

According to local characteristics of the breeding plants, the fully-mixed medium-temperature two-stage fermentation technology is selected. On the one hand, 20 tons of pig manure and 20 m^3 of washing waste water are fermented by using a CSTR. The daily biogas production is 1200 m^3, which meets the gas demands of nearby 828 households for three meals a day and the plants. The remaining about 30 tons of pig manures are deeply treated in the organic fertilizer plant and sold. Biogas residue and slurry are not separated but directly used as fertilizer and applied to surrounding farmlands. On the other hand, the remaining 250 m^3 of washing waste water from the breeding plant is precipitated and then put into the secondary fermentation pool and finally used as biogas slurry fertilizer after undergoing fully-mixed anaerobic treatment, which meets the fertilizer demands of surrounding farmlands. This project uses moderate-temperature two-stage anaerobic fermentation process and two-stage anaerobic, integrated, double-layer flexible membrane gas holder. The project of centralized biogas supply completed in 2009 actually provides daily biogas for cooking of over 1200 households in the village and neighboring villages and gas transmission pipelines are up to 14 km, with the furthest end reaching 4 km away (Figs. 5.8 and 5.9). At that time, it was the largest centralized village-level gas supply project in China.

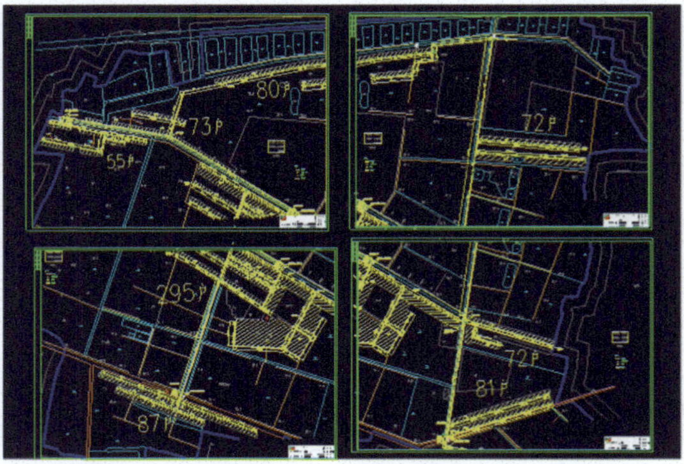

Fig. 5.8 Design of pipelines

Fig. 5.9 Biogas transmitted into households

The technology used in the breeding plant for constructing ecological pig farm with resourceful utilization of pollutants from breeding belongs to cleaning production technology. Before constructing biogas project, sewages and livestock manure discharged from the breeding plant resulted in great pollutions to environment and seriously polluted soils, air and water body around the breeding plant. In particular, groundwater pollution seriously affected people's health. After constructing biogas station, these pollutants are transformed into energy through anaerobic fermentation. In the fermentation, germs are killed and pollution reduces. The remaining biogas residues and slurry are applied as fertilizers to the farmlands, which can minimize the pollution to the environment and achieve good environmental benefits. In addition, biogas, as a clean energy is used by all villagers for daily life, thus producing social benefits.

Chapter 6
Conclusions

Organic waste is an important factor polluting urban and rural environment in China and influencing village appearance and residents' health in villages and towns. By taking organic wastes produced from living, decentralized breeding industries and service industry in villages and towns as objects, the research on differentiated resourceful utilization of organic wastes in villages and towns was carried out. This book mainly performed the research on key process technologies, such as fertilization and energy regeneration suitable for being promoted in villages and towns and developed a complete set of diversified, differentiated, high-value, low-pollution treatment and disposal equipments for organic wastes in villages and towns. The practical equipment for treating organic wastes in villages and towns was used. Based on this, the technical mode of manageable techniques and equipment of low pollution, low cost, and low maintenance requirement suiting for resourceful utilization of organic wastes in villages and towns was built by integrating mature and well-matched resourceful technologies of organic wastes. The relevant technical standards and specifications were compiled and long-term operation mechanisms were established. This fundamentally raises the capacity and level of resourceful and harmless treatment of organic wastes in villages and towns in China, improves the environment quality in villages and towns and promotes sustainable development of urbanization construction.

1. In disposal of organic wastes from decentralized breeding in villages and towns, the problems existing in resourceful treatment technologies were analyzed by studying generation characteristics of these organic wastes in villages and towns in China. The bio-augmented composting technology and associated functional bacteria agents of organic wastes were mainly developed to improve the efficiency of aerobic biological fermentation of organic wastes and the quality of resourceful products. In view of problems of low biogas production rate, instability and difficulties in feeding and discharge in anaerobic fermentation, key technologies and equipments for high-efficient anaerobic fermentation of livestock manure in decentralized breeding were developed. In the meanwhile, environmental safety evaluation and fertilizer efficiency of resourceful utilization through composting

M. Li et al., *Differentiated Resourceful Utilization of Rural Organic Wastes*,
https://doi.org/10.1007/978-981-15-2712-8_6

of derived products from resourceful treatment of organic wastes, such as biogas residues in villages and towns were carried out. By integrating key technologies, multi-level key technologies and equipments of resourceful utilization via anaerobic fermentation of decentralized breeding wastes in villages and towns and fertilizer production through aerobic fermentation with stirred biogas residues were integrated.

2. As for utilization of mixed organic wastes from pedlars' markets in villages and towns, the wastes were classified into green wastes with high cellulose, wastes of perishable stems and leaves and mixed wastes based on the research and development of differentiated classification technology of these organic wastes. Furthermore, this book built a low-cost and high-efficiency management mode suiting for differentiated classification and resourceful utilization of organic wastes from pedlars' markets in villages and towns. Wastes with high content of lignocelluloses have characteristics, such as low heat value, high water content and poor biodegradability. Aiming at this, based on the research and development of high-efficient pre-treatment technology and bacteria agent for degrading lignocellulose, this book mainly developed the multi-stage inoculation technology for preparing high-quality humic acid fertilizer by using bio-augmented technology. In view of characteristics of high content and complex compositions of organic matter of mixed wastes of stems and leaves and animal residues, the problems of instable operation and low biogas production of anaerobic fermentation were mainly solved. By optimizing process parameters, the CHMP-AF technology was researched and developed. By integrating key technologies, the key technologies and equipments for differentiated resourceful utilization of organic wastes from pedlars' markets in villages and towns were developed.
3. The high water content, high oil content and high salt content are not conducive to biological fermentation of kitchen wastes. In view of this problem and to realize the resourceful treatment, hydrothermal pre-treatment technologies and equipments were developed for kitchen wastes. The bio-augmented composting technology and associated equipments of organic wastes as well as bacteria agent with the anti-acidification function were developed to improve aerobic fertilizer efficiency of kitchen wastes and the quality of recycled products. In addition, the high-efficient anaerobic fermentation and co-production technology of kitchen wastes was researched. This book optimized solid-liquid coupling parameters, reduced emission of biogas slurry and developed resourceful utilization technology of biogas residues through bio-augmented composting. The resourceful treatment technologies and equipments of kitchen wastes integrating pre-treatment, bio-augmented inoculation and biological fermentation were formed. Developing ecological agriculture can realize the virtuous cycle of kitchen wastes deriving from farmland, produced on table, and finally returned to farmland.

The manufacturer's authorised representative in the EU is Springer Nature Customer Service Centre GmbH, Europaplatz 3, 69115 Heidelberg, Germany. If you have any concerns regarding our products, please contact ProductSafety@springernature.com

Printed and bound by CPI Group (UK) Ltd, Croydon, CR0 4YY
30/09/2026
02225574-0009